THE HAND ATLAS

THE HAND ATLAS

By

MOULTON K. JOHNSON, M.D.

Chief, Hand Clinic
Associate Clinical Professor of Surgery/Orthopedics
U.C.L.A. School of Medicine
Los Angeles, California
Senior Surgeon, Santa Monica Hospital
Senior Surgeon, St. John's Hospital
Santa Monica, California

and

MYLES J. COHEN, M.D.

Department of Surgery/Orthopedics
U.C.L.A. School of Medicine
Los Angeles, California

CHARLES C THOMAS · PUBLISHER
Springfield · Illinois · U.S.A.

© 1975, by CHARLES C THOMAS · PUBLISHER

ISBN 0-398-03203-3

Library of Congress Catalog Card Number: 74-13144

Printed in the United States of America

P-4

Library of Congress Cataloging in Publication Data

Johnson, Moulton K
    The hand atlas.

    Bibliography: p.
    1. Hand—Atlases.   2. Anatomy, Surgical and topographical.
I. Cohen, Myles J., joint author.   II. Title.
[DNLM: 1. Hand—Anatomy and histology—Atlases WE17 J68h]
QM548.J57   611'.91'0222   74-13144
ISBN 0-398-03203-3

TO

CITA AND GINNY,
who don't complain.

INTRODUCTION

THIS ATLAS ATTEMPTS TO PORTRAY the anatomy of the hand as the surgeon will encounter it in the operating room. The surgeon normally makes an incision in the skin and works his way down to the pathology, layer by layer. Therefore, *The Hand Atlas* depicts the anatomy layer by layer, beginning with the skin and ending with X-ray studies of the bones.

Most anatomy texts are illustrated with drawings. The drawings are usually rendered skillfully, but often the details are distorted by the artist's conception of the anatomy, by his imagination, or by the difficulties inherent in depicting textures and complex shapes. At the operating table the surgeon may be confused and disconcerted when he is unable to find structures which closely resemble the drawings in the text. The problem is compounded when the drawings are made from specimens which have been distorted by the embalming process. *The Hand Atlas*, therefore, is illustrated with photographs of dissections of unembalmed specimens. Hopefully, the authors have provided more realistic illustrations by substituting the surgeon's camera for the artist's brush.

In the studies of the arterial circulation the vessels have been injected with red latex. Similarly, in some instances the tendon sheaths have been injected with black silastic. To demonstrate the potential size of the major spaces of the hand, they have been injected with radiopaque contrast media. All illustrations are approximately life-size except where otherwise indicated. With these exceptions, the anatomical structures illustrated are as the authors found them and as the surgeon may logically expect to encounter them. It cannot be overemphasized, however, that anatomical variations are both numerous and common in the hand.

As *The Hand Atlas* is intended as a practical guide for the surgeon, the anthropology and embryology of the hand are beyond its scope. Any reader interested in these areas is referred to Dr. Emanuel Kaplan's classic work, *Functional and Surgical Anatomy of the Hand*, which contains a wealth of detailed information which the atlas format does not permit us to include. The authors have made liberal use of Dr. Kaplan's book in the preparation of this atlas.

Throughout *The Hand Atlas* the authors have used as a guide *Terminology for Hand Surgery*, published by the International Federation of Societies for Surgery of the Hand in 1970.

ACKNOWLEDGMENT

We gratefully acknowledge the invaluable editorial assistance of Neuropod E. Yuhl, M.D., who spent many hours clarifying the text.

M.K.J.
M.J.C.

CONTENTS

THE HAND ATLAS

CHAPTER I

DISTAL VOLAR FOREARM

THE DISTAL THIRD OF THE VOLAR ASPECT of the forearm contains a number of important anatomical structures. Included among these structures are nerves, arteries and tendons, as well as the overlying skin, fat and fascia. For descriptive purposes within this text, the authors have arbitrarily divided this area of the forearm into three layers.

The superficial layer is comprised of the skin, the subcutaneous fat, the tendon of the palmaris longus, and the antebrachial fascia. The intermediate layer includes the radial artery, the flexor carpi radialis, the median nerve, the flexors digitorum superficialis, and the flexor carpi ulnaris. The flexor pollicis longus, the flexors digitorum profundus and the ulnar nerve, artery and veins comprise the deep layer.

SUPERFICIAL LAYER

The skin of the volar surface of the forearm is very thin. The subcutaneous fat layer is proportionately thin in lean individuals. The tendon of the palmaris longus (Fig. 1) is the most superficial structure beneath the subcutaneous fat. This tendon (present in about 90% of individuals*) lies superficial to the antebrachial fascia. The palmaris longus inserts into the apex of the palmar aponeurosis at the proximal border of the palm. The thin, translucent antebrachial fascia blends distally into the thicker flexor retinaculum.

---

* Because of the human tendency to see what one expects to see, more than one unwary surgeon has excised the median nerve to use for a free tendon graft, mistaking the nerve for an absent palmaris longus tendon.

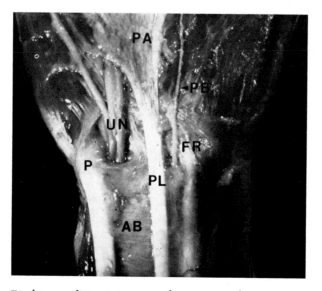

Figure 1. Palmaris longus. Pisohamate ligament removed, exposing ulnar nerve in Guyon's canal. Note that in this specimen, palmar cutaneous branch of median nerve is superficial to flexor retinaculum and to origins of thenar muscles. AB = antebrachial fascia; FR = flexor retinaculum; P = pisiform; PA = palmar aponeurosis; PB = palmar cutaneous branch of median nerve; PL = palmaris longus; UN = ulnar nerve.

The Hand Atlas

4

INTERMEDIATE LAYER
Radial Artery

The radial artery is situated most laterally of all the structures of the intermediate layer. The artery passes distally along the lateral edge of the radius. At the styloid process of the radius, the radial artery gives off a superficial palmar branch. This branch enters the palm passing either over the palmar surface of the abductor pollicis brevis (Fig. 11) or through that muscle (Fig. 13).

After giving rise to the superficial palmar branch, the radial artery turns dorsally. It passes distally beneath the tendons of the abductor pollicis longus,

and the extensors pollicis longus and brevis (Fig. 63). Emerging from beneath these tendons, the radial artery courses over the dorsal surface of the first dorsal interosseous. (The dorsal branches of the radial atrery are described in Ch. VI.) The artery then passes between the two heads of the first dorsal interosseous into the palm and anastamoses with the deep branch of the ulnar artery to form the deep palmar arch (Ch. II).

Flexor Carpi Radialis

The large tendon of the flexor carpi radialis lies adjacent to and medial to the radial artery (Fig. 2). It passes distally through a separate tunnel

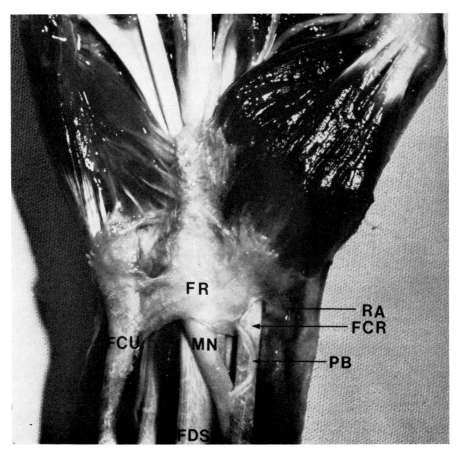

Figure 2. Flexor retinaculum. Palmaris longus, antebrachial fascia and flexor tendon shreaths removed. Note that in this specimen palmar cutaneous branch of median nerve penetrates substance of flexor retinaculum and muscle bellies of flexors digitorum superficialis extend into carpal tunnel. FCR = flexor carpi radialis; FCU = flexor carpi ulnaris; FDS = flexor digitorum superficialis to long finger; FR = flexor retinaculum; MN = median nerve; PB = palmar cutaneous branch of median nerve; RA = radial artery.

over the volar surface of the carpal bones into the hand. The main insertion of the flexor carpi radialis is into the second metacarpal. This powerful wrist flexor is innervated by the median nerve.

Median Nerve

The median nerve emerges in the distal forearm from beneath the muscle bellies of the flexors digitorum superficialis. The nerve follows an oblique course medially and becomes superficial to the tendons of the flexors digitorum superficialis but remains deep to the palmaris longus and to the antebrachial fascia (Fig. 1)† ‡.

Flexor Digitorum Superficialis

The flexors digitorum superficialis occupy the space between the flexor carpi radialis and the flexor carpi ulnaris. The bellies of the flexors digitorum superficialis extend distally for varying distances. Not infrequently one or more muscle bellies will encroach on the carpal tunnel (Fig. 2). The superficialis tendons to the long and ring fingers are superficial to those of the index and little fingers. Inserting into the middle phalanges of the four fingers (Ch. IV), the flexors digitorum superficialis are strong flexors of the proximal interphalangeal joints, but they are not efficient flexors of the metacarpophalangeal joints because they pass close to the axis of flexion of these joints. The flexors digitorum superficialis are innervated by the median nerve.

Flexor Carpi Ulnaris

The flexor carpi ulnaris is the most medially situated of the structures of the intermediate layer of the distal third of the volar forearm. The tendon inserts primarily into the pisiform, although some fibers blend into the flexor retinaculum as well as into the pisohamate ligament (Figs. 1, 2). The flexor carpi ulnaris acts as an effective flexor as well as an ulnar deviator of the wrist joints. The muscle is supplied by the ulnar nerve.

DEEP LAYER
Ulnar Nerve and Ulnar Vessels

The ulnar artery, its two accompanying veins, and the ulnar nerve are in close juxtaposition beneath the flexor carpi ulnaris. The nerve and the vessels pass between the pisiform and the hook

† The palmar branch of the median nerve comes off the lateral side of the nerve a few centimeters proximal to the wrist. This branch may pass distally superficial to the flexor retinaculum (Fig. 1) or may penetrate the substance of the retinaculum (Fig. 2) to emerge distally in the subcutaneous tissues of the palm. The palmar branch of the median nerve supplies the skin of the radial half of the proximal palm.

‡ During decompression of the median nerve by division of the flexor retinaculum, the surgeon should stay to the ulnar side of the median nerve. By so doing, he will protect both the palmar cutaneous and the recurrent motor branches of the median nerve.

of the hamate in a ligament-covered structure named Guyon's canal (Figs. 12, 79). The course of the ulnar nerve and artery distal to Guyon's canal is described in Chapter II.

Flexor Digitorum Profundus

The tendons of the flexors digitorum profundus are deep to the tendons of the flexors digitorum superficialis and are superficial to the pronator quadratus. The profundus tendons pass distally through the carpal tunnel and extend out on the fingers to insert into the distal phalanges. The flexors digitorum profundus are flexors of both the proximal and distal interphalangeal joints of the fingers. The tendons pass close to the axis of flexion of the metacarpophalangeal joints and are therefore inefficient in flexing these joints. The profundi of the ring and little fingers are supplied by the ulnar nerve. The median nerve supplies the profundi of the index and long fingers.

Flexor Pollicis Longus

The flexor pollicis longus lies deep to the flexor carpi radialis and lateral to the flexors digitorum profundus. The flexor policis tendon courses distally through the carpal tunnel, passes beneath the superficial thenar muscles and thereafter inserts into the distal phalanx of the thumb. The flexor pollicis longus is a strong flexor of the interphalangeal joint of the thumb, but it is inefficient as a flexor of the metacarpophalangeal joint. The muscle is supplied by the median nerve.

CARPAL TUNNEL

The carpal tunnel begins at the distal wrist flexion crease. This tunnel provides access to the palm for the extrinsic flexor tendons of the digits and for the median nerve. Viewed from the palmar side of the hand, the lunate and the capitate comprise the floor of the tunnel. The tubercle of the trapezium forms the wall on the lateral side. The hook of the hamate forms the medial wall (Figs. 5, 6). The flexor retinaculum is the roof.

Flexor Retinaculum

The flexor retinaculum is a strong, inelastic structure several millimeters in thickness (Figs. 3 and 4). Proximally, it is continuous with the antebrachial fascia. Distally the retinaculum thins rapidly in the midpalm and merges into a tenuous layer of transparent fascia deep to the palmar aponeurosis. The flexor retinaculum is firmly attached to the hook of the hamate and to the pisiform on the medial side of the hand and to the tubercles of the trapezium and of the scaphoid on the lateral side of the hand.

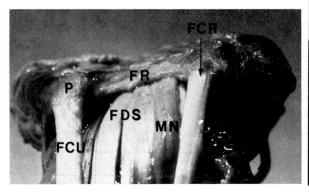

Figure 3. Flexor retinaculum, tangential view. Wrist in hyperextension. Note thickness of flexor retinaculum. FCR = flexor carpi radialis; FCU = flexor carpi ulnaris; FDS = flexors digitorum superficialis; FR = flexor retinaculum; MN = median nerve; P = pisiform.

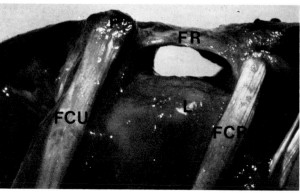

Figure 4. Carpal tunnel. Median nerve and digital flexor tendons and sheaths removed. Hyperextended wrist is viewed tangentially, looking distally from lower forearm. FCR = flexor carpi radialis; FCU = flexor carpi ulnaris; FR = flexor retinaculum; L = lunate; P = pisiform.

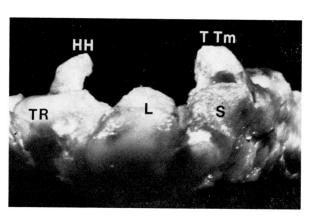

Figure 5. Osseous framework of carpal tunnel. Flexor retinaculum, muscles, tendons and pisiform removed. Specimen is viewed tangentially as in Figure 4. HH = hook of hamate; L = lunate; S = scaphoid; TTm = tubercle of the trapezium.

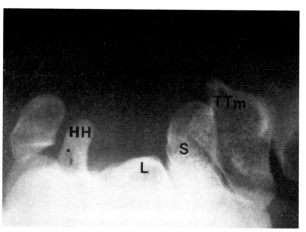

Figure 6. X-ray view of the carpal tunnel. HH = hook of the hamate; L = lunate; S = scaphoid; TTm = tubercle of the trapezium.

CHAPTER II

THE PALM

THE PALM IS THE MOST COMPLEX area of the hand. In the central palm eight different layers may be identified. The three most superficial layers are bound closely together. They are the skin, the subcutaneous fat, and the palmar aponeurosis. The palm contains two arterial arches. The superficial palmar arch lies beneath the palmar aponeurosis. The arch is superficial to the sensory branches of the median and ulnar nerves, which in turn are superficial to the flexor tendons and to the lumbrical muscles. The deep palmar arch with the deep motor branch of the ulnar nerve comprises the next layer. The intrinsic muscles and the metacarpals constitute the bottom layer.

Skin and Subcutaneous Fat

The skin of the palm is highly specialized. The palmar skin is sharply demarcated from the skin of the forearm by the most distal wrist flexion crease (Fig. 7). The palmar skin has a thick cornified layer and is relatively inelastic. It is well endowed with sensory nerve endings and with sweat glands. Papillary ridges increase its surface area. The palmar skin is minimally pigmented, even in the dark-skinned races.

The underlying mobility of the hand is manifested in the thick palmar skin by the skin creases. The most prominent are the thenar creases, the distal and proximal palmar creases, and the median crease. The crease pattern varies somewhat from person to person, but it bears a relatively constant relationship to the underlying structures (Figs. 8, 9, 10).

The layer of subcutaneous fat overlying the thenar muscles may be as thin as a millimeter or two in lean individuals. In obese subjects the layer over the heavily padded metacarpal heads may exceed a centimeter in thickness. The fat layer is much thinner at the creases. Fibrous bands tether the skin and fat to the underlying fascial and ligamentous structures.

Palmar Aponeurosis

The palmar aponeurosis is triangular in shape and covers the central and distal palm (Fig. 11). It is composed of a superficial layer of longitudinal fibers and a deeper layer of transverse fibers. The aponeurosis begins at the center of the proximal border of the flexor retinaculum. Whenever the palmaris longus muscle is present, its tendon inserts into the palmar aponeurosis. The longitudinal fibers divide in the proximal palm into four bundles which extend distally. These bundles overlay the flexor tendons of the four fingers and are therefore called pretendinous bands. The pretendinous bands insert firmly into the sheaths of the flexor tendons at the metacarpophalangeal joints.

The transverse fibers of the palmar aponeurosis fasten the pretendinous bands together, primarily by attachments on their deep surfaces. The distal portion of the transverse fibers is the more prominent and is called the superficial transverse palmar ligament. The transverse fibers do not extend distally beyond the metacarpophalangeal joints.

Distal and parallel to the superficial transverse palmar ligament lies the natatory ligament (Fig. 11). This ligament is attached to the volar surfaces of the flexor sheaths at the metacarpophalangeal joint level and forms the floors of the lumbrical canals. The lumbrical muscles, the digital nerves and the digital arteries pass through these canals.

Palmaris Brevis

The palmaris brevis is of variable size and of dubious significance (Fig. 20). When relatively large, the muscle overlies much of the proximal portion of the hypothenar fascia and inserts into the proximal apex of the palmar aponeurosis. At

7

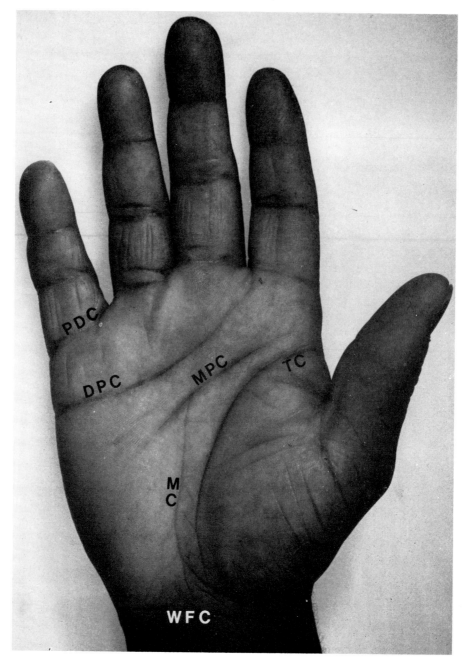

Figure 7. Palmar skin creases. DPC = distal palmar crease; MC = median crease; PDC = palmar digital crease; MPC = proximal palmar crease; TC = thenar crease; WFC = wrist creases.

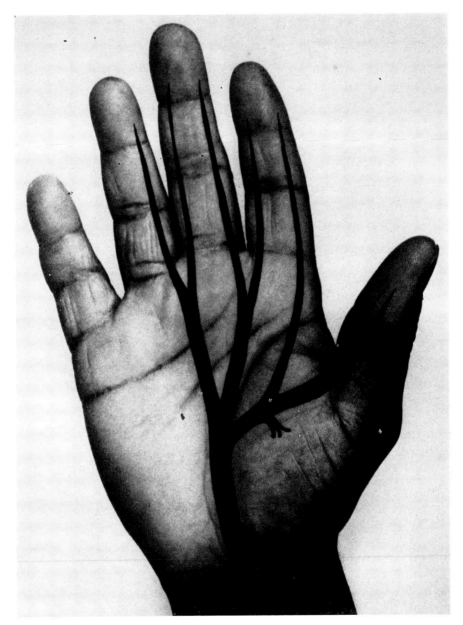

Figure 8. Median nerve. Course of the median nerve shown schematically in relation to surface topography.

The Hand Atlas

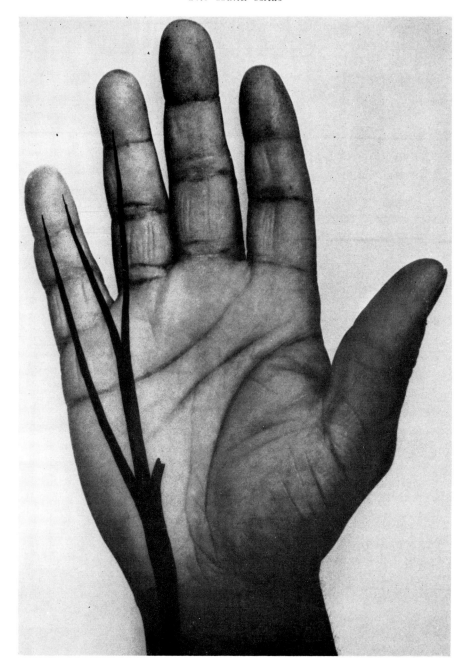

Figure 9. Ulnar nerve. Course of ulnar nerve shown schematically in relation to surface topography.

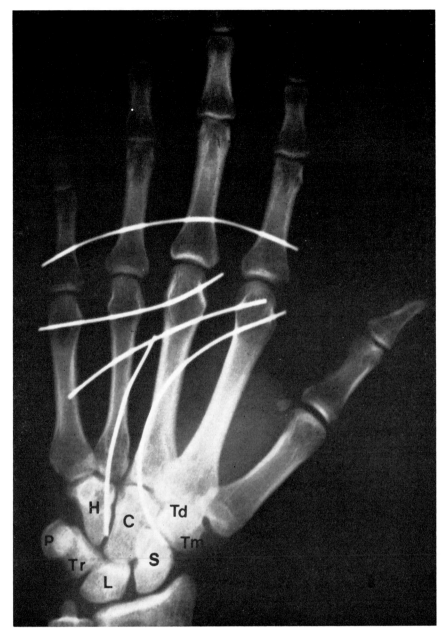

Figure 10. Anteroposterior X-ray view of hand with location of palmar skin creases superimposed (white lines). C = capitate; H = hamate; L = lunate; P = pisiform; S = scaphoid; Td = trapezoid; Tm = trapezium; Tr = triangular.

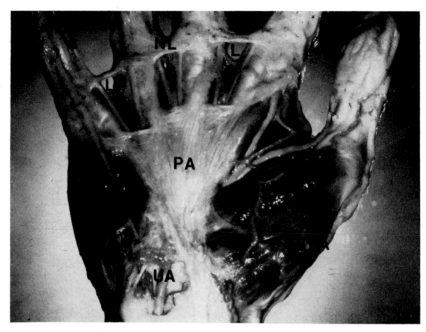

Figure 11. Palmar aponeurosis and natatory ligament. Skin, subcutaneous fat, thenar and hypothenar fascia removed. Note that in this specimen the superficial branch of the radial artery passes superficially over abductor pollicis brevis and flexor pollicis brevis muscles. L = lumbrical canal; NL = natatory ligament; PA = palmar aponeurosis; PP = princeps pollicis artery; SBRA = superficial branch of radial artery; UA = ulnar artery.

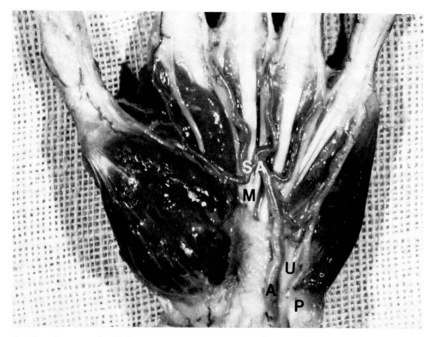

Figure 12. Superficial palmar arch. Palmar aponeurosis removed. Note that superficial palmar arch is superficial to branches of median and ulnar nerves, as well as to flexor tendons and lumbrical muscles. A = ulnar artery; M = median nerve; P = pisiform; SA = superficial arch; U = ulnar nerve.

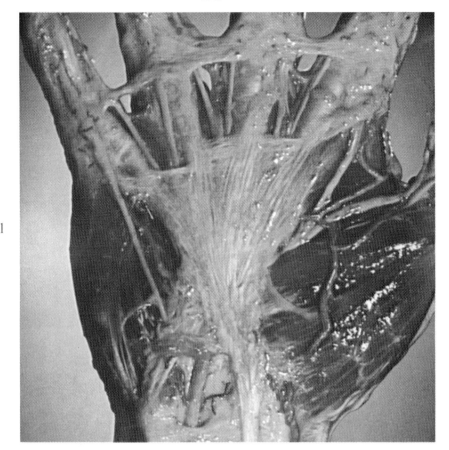

Plate 1

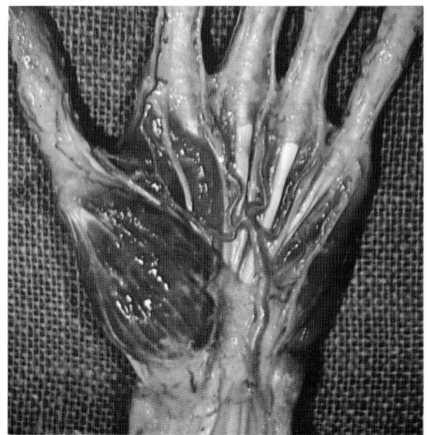

Plate 2

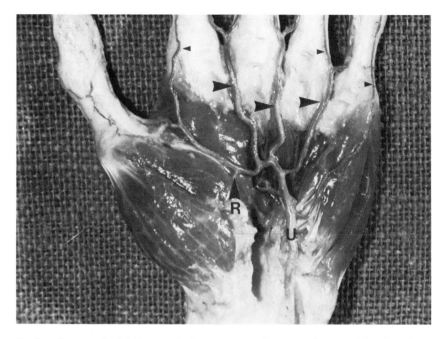

Figure 13. Superficial palmar arch. Median and ulnar nerves, flexor tendons and lumbrical muscles, and flexor retinaculum removed. Note common palmar digital arteries (large arrows) and proper palmar digital arteries (small arrows). SA = superficial arch; R = radial artery; U = ulnar artery.

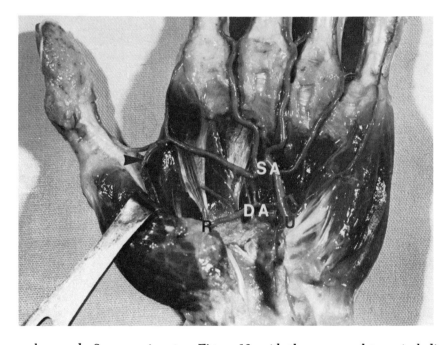

Figure 14. Deep palmar arch. Same speciment as Figure 13, with thenar musculature, including the adductor pollicis, retracted. Note contributions of radial artery (arrow) to superficial arch. DA = deep arch; RA = radial artery; SA = superficial arch; UA = ulnar artery.

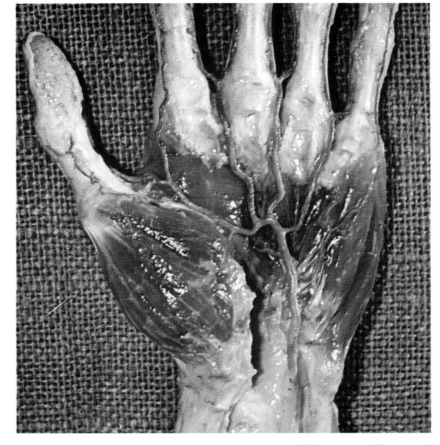

Plate 3

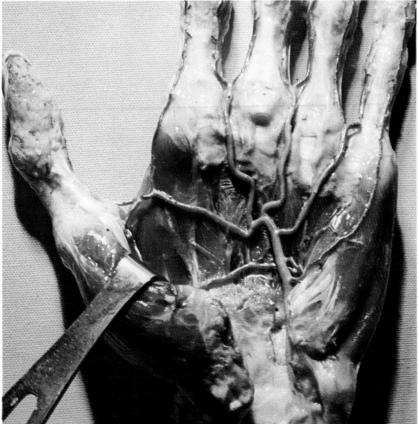

Plate 4

The Hand Atlas

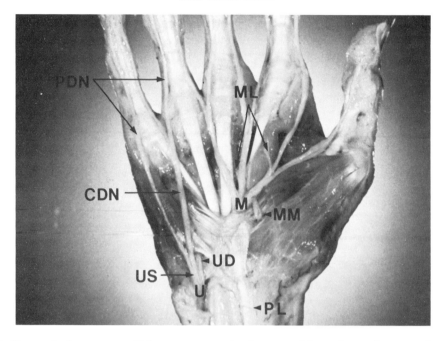

Figure 15. Median and ulnar nerves. Palmar aponeurosis, thenar and hypothenar fascia, and arteries removed. CDN = common palmar digital nerve; M = median nerve; ML = median motor branches to first and second lumbrical muscles; MM = median motor branch; PDN = proper palmar digital nerve; PL = palmaris longus; U = ulnar nerve; UD = ulnar nerve deep motor branch; US = ulnar nerve superficial branch.

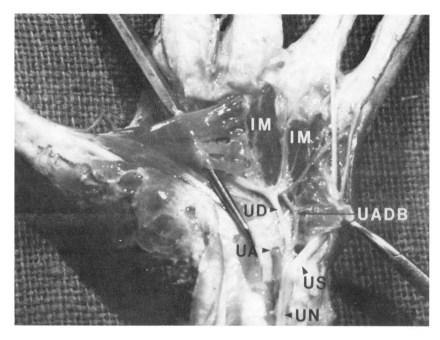

Figure 16. Ulnar nerve. Flexor tendons, superficial arch and median nerve removed, origins of hypothenar muscles pulled aside by hemostat, probe lies between transverse and oblique heads of adductor pollicis. IM = interosseous muscles; UA = ulnar artery; UADB = ulnar artery deep branch; UD = ulnar nerve deep motor branch; UN = ulnar nerve; US = ulnar superficial branch.

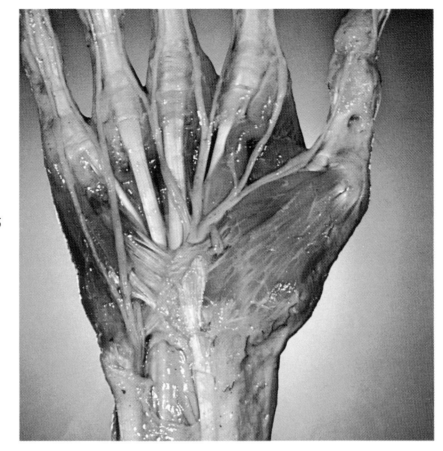

Plate 5

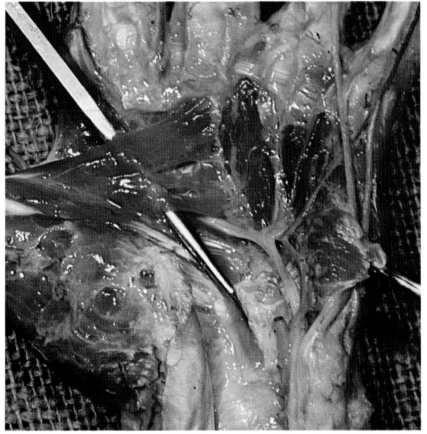

Plate 6

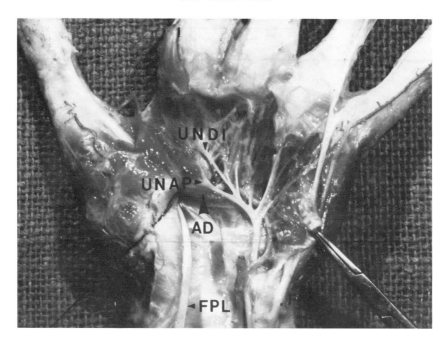

Figure 17. Ulnar nerve. Same specimen as Figure 16, with origin of transverse head of adductor pollicis detached from third metacarpal and retracted radially. AD = origin of oblique head of adductor pollicis; FPL = flexor pollicis longus; UNAP = ulnar nerve branch to adductor pollicis; UNDI = ulnar nerve branch to first dorsal interosseous.

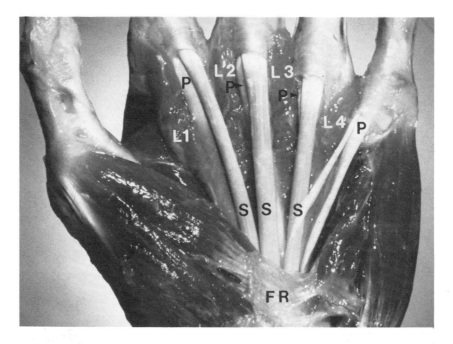

Figure 19. Flexor tendons. Palmar aponeurosis, thenar and hypothenar fascia, nerves and vessels removed. In this specimen flexor digitorum superficialis to little finger is a slender slip from flexor superficialis to ring finger. FR = flexor retinaculum; L1,2,3,4 = lumbricals 1,2,3,4; P = flexor digitorum profundus; S = flexor digitorum superficialis.

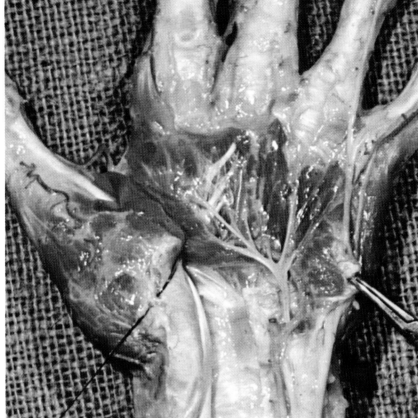

Plate 7

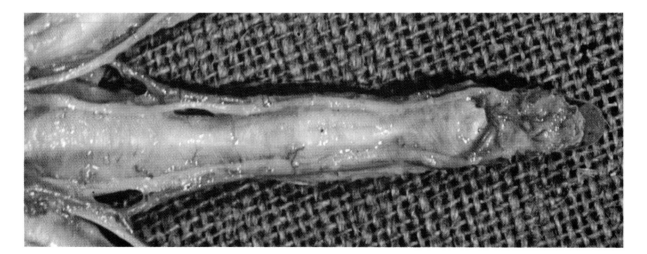

Plate 8 (Figure 18). Proper palmar digital nerves and arteries of the finger. Arteries injected with red latex.

other times the palmaris brevis consists only of a few vestigial muscle fibers. It is innervated by the ulnar nerve.

ARTERIES OF THE PALM

The two usual sources of arterial supply of the palm are the radial artery and the ulnar artery.[1] In some 5 per cent of cases, a large median nerve artery contributes. The palm and the volar portion of the digits are supplied by two arterial arches, the superficial and deep palmar arches.

Superficial Palmar Arch

The superficial palmar arch traverses the palm beneath the palmar aponeurosis superficial to the sensory branches of the median and ulnar nerves (Fig. 12). The arch is formed by the superficial branch of the ulnar artery approximately two thirds of the time. In about a third of specimens, the superficial palmar branch of the radial artery anastamoses with the branch of the ulnar artery to form the superficial arch (Figs, 11, 13). In perhaps 5 per cent of specimens the median nerve artery is unusually large. In those instances the median nerve artery, which arises from varying sources in the proximal forearm, forms the superficial arch, with or without contributions from the radial and ulnar arteries.

Common Palmer Digital Arteries

The common palmar digital arteries arise from the superficial palmar arch. Most often there are three common palmar digital arteries. They bifurcate proximal to the interdigital folds, forming proper digital arteries for the adjacent surfaces of the index, long, ring and little fingers. The proper palmar digital artery to the ulnar side of the little finger comes directly off the superficial arch.

Deep Palmar Arch

The deep palmar arch courses across the palm beneath the tendons of the flexors digitorum profundus and the lumbrical muscles. The arch is formed by anastamosis of the termination of the radial artery with the deep branch of the ulnar artery (Fig. 14).

Palmar Metacarpal Arteries

The deep palmar arch gives off the princeps

[1] Sherman S. Coleman and Barry J. Anson, "Arterial Patterns in the Hand Based Upon a Study of 650 Specimens," *Surgery, Gynecology and Obstetrics* (Chicago), Vol. 113 (1961), p. 409.

pollicis artery to the thumb and usually three palmar metacarpal arteries. The palmar metacarpal arteries run a course parallel to the larger common palmar digital arteries. There are numerous but inconstant anastamoses between the palmar metacarpal arteries and the dorsal metacarpal arteries and between the palmar metacarpal arteries and the common palmar digital arteries at their bifurcations.

The variations of the arteries of the palm are legion. The superficial arch is more variable than the deep arch. When a common palmar digital artery is lacking or is quite small, the deficit is compensated for by increased size of the corresponding palmar metacarpal artery. Occasionally a dorsal metacarpal artery may be unusually large and be the main supplier of a pair of proper palmar digital arteries.

MEDIAN NERVE

The median nerve enters the palm between the flexor retinaculum and the tendons of the flexors digitorum superficialis. The palmar cutaneous branch of the median nerve enters the palm separately (Ch. I). This nerve supplies the skin of the radial side of the proximal palm.

The median nerve usually branches at the distal margin of the flexor retinaculum. The pattern of arborization varies. Most often the nerve bifurcates into a lateral and a median branch (Fig. 15).

The medial branch promptly divides again into the second and third common palmar digital nerves. The common digital nerves in turn bifurcate at the level of the metacarpophalangeal joints to form the proper palmar digital nerves to the adjacent surfaces of the index, long, and ring fingers. The second common digital nerve gives off a small motor branch to the second lumbrical muscle before dividing into proper digital nerves.

The lateral branch of the median nerve immediately gives off the recurrent motor branch. This branch angles back in a proximal radial direction for a few millimeters, disappears beneath the abductor pollicis brevis, and arborizes into terminal twigs. These twigs supply the abductor pollicis brevis, the opponens pollicis, and the superficial head of the flexor pollicis brevis.

The lateral branch of the median nerve most often divides into a first common digital nerve and radial proper digital nerve to the thumb. The first common digital nerve gives off a motor twig to the first lumbrical before bifurcating into the radial proper digital nerve of the index and ulnar proper digital nerve of the thumb. The radial proper digital nerve to the thumb crosses the flexor pollicis longus tendon proximal to the metacarpophalangeal joint.

In most hands the median nerve supplies the volar aspect of the thumb, the index and long fingers, and of the radial half of the ring finger. The ulnar nerve supplies the volar aspect of the ulnar half of the ring finger and of both sides of the little finger. However, exceptions to this pattern occur occasionally. In such instances the median nerve may serve both halves of the volar aspect of the ring finger, or conversely the ulnar nerve may supply the adjacent surfaces of the long and ring fingers. Anastomosis between the median and ulnar nerves may occur both proximal to the wrist and in the palm. Figure 19 illustrates such an anastomosis in the palm.

ULNAR NERVE

The ulnar nerve enters the palm through Guyon's canal. It divides into a deep and a superficial branch after emerging from Guyon's canal. The superficial sensory branch divides into the ulnar proper digital nerve of the little finger and the fourth common digital nerve. Occasionally the ulnar proper digital nerve of the little finger leaves

the main ulnar trunk proximal to the wrist and courses over the surface of the abductor digiti minimi to reach the little finger. The fourth common digital nerve in turn bifurcates into proper digital nerves supplying the adjacent surfaces of the ring and little fingers.

The deep branch of the ulnar nerve is a motor nerve. It passes between, and innervates, the abductor digiti minimi and the flexor digiti minimi, and supplies as well the opponens digiti minimi. Thereafter, the motor branch descends deep into the palm along with the deep branch of the ulnar artery (Fig. 16). In the proximal palm, the nerve curves laterally, providing small branches to the third and fourth lumbricals, and in turn to each interosseous muscle. The terminal branches of the deep branch of the ulnar nerve pass between the transverse and oblique heads of the adductor pollicis to supply both the adductor and the deep head of the flexor pollicis brevis (Fig. 17).

FLEXOR TENDONS

The tendons of the flexors digitorum superficialis and flexors digitorum profundus enter the palm through the carpal tunnel. As the tendons emerge beneath the distal border of the flexor retinaculum, they fan out toward the fingers (Fig. 18). The tendons lie deep to the median nerve, to the sensory branches of the ulnar nerve, and to the super-

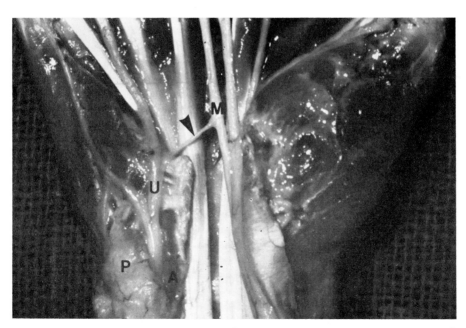

Figure 19. Abnormal connections between median and ulnar nerves. Flexor retinaculum opened and superficial arch removed. Arrow = connecting nerve branch; A = ulnar artery emerging (with ulnar nerve) from Guyon's canal; M = common palmar digital nerve from median nerve; P = pisiform; U = ulnar nerve.

ficial palmar arch (Fig. 12). The deep branch of the ulnar nerve and the deep palmar arch cross the palm beneath the flexor tendons. The tendons of the flexors digitorum profundus lie deep to those of the flexor digitorum superficialis. The lumbrical muscles arise from the profundus tendons in the proximal palm (Fig. 18). Each lumbrical passes through the lumbrical canal on the radial side of the appropriate finger. The flexor tendons leave the palm and pass through the fibrous sheaths of the fingers which begin at the metacarpophalangeal joints.

THENAR MUSCLES

The thenar muscles are the abductor pollicis brevis, the opponens pollicis, the flexor pollicis brevis, and the adductor pollicis. They are covered on their volar surface by the thin thenar fascia, which is continuous with the palmar aponeurosis.

Abductor Pollicis Brevis

The abductor pollicis brevis, the most superficial, arises from the radial half and the distal edge of the flexor retinaculum (Figs. 20, 21). Its origin frequently includes a slip of tendon from the abductor pollicis longus. The muscle inserts mainly by a distinct tendon into the lateral capsule of the metacarpophalangeal joint of the thumb (Fig. 37). The abductor pollicis brevis abducts and flexes the first metacarpal and rotates both the thumb and the first metacarpal. This muscle alone can produce nearly complete opposition. It

is also an extensor of the interphalangeal joint because it contributes fibers to the extensor mechanism. The muscle is supplied by the median nerve.

Opponens Pollicis

The opponens pollicis originates from the flexor retinaculum and from the tubercle of the trapezium and inserts along the length of the radial border of the first metacarpal (Fig. 22). The muscle is a flexor and rotator of the first metacarpal, thereby assisting the abductor pollicis brevis in opposing the thumb. It is innervated by the median nerve.

Flexor Pollicis Brevis

The flexor pollicis brevis is composed of two heads (Fig. 23). The superficial head is innervated by the median nerve and the deep by the ulnar nerve. The superficial head arises from the distal radial border of the flexor retinaculum and from the tubercle of the trapezium. The deep head originates from the palmar surface of the trapezoid and the adjacent portion of the capitate. The deep head varies greatly in size, constituting from 10 to 40 per cent of the muscle mass. The two heads insert by a common tendon into the radial sesamoid of the metacarpophalangeal joint of the thumb. Secondary slips insert into the lateral tubercle at the base of the proximal phalanx and into the extensor mechanism. The primary function of this muscle is to flex the metacarpal and the metacarpophalangeal joint of the thumb. As the flexor pollicis brevis

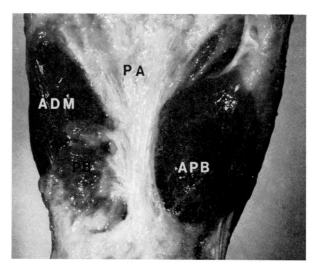

Figure 20. Palmaris brevis. This muscle is rarely larger and often smaller than shown here. ADM = abductor digiti minimi; APB = abductor pollicis brevis; PA = palmar aponeurosis; PB = palmaris brevis.

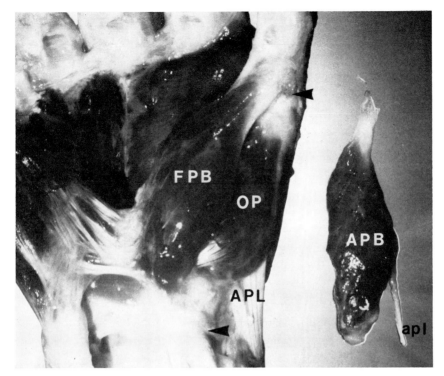

Figure 21. Abductor pollicis brevis. Palmar aponeurosis, nerves, vessels and extrinsic flexor tendons to thumb and fingers removed. Abductor pollicis brevis detached from origin on flexor retinaculum (lower arrow) and from insertion (upper arrow), exposing opponens pollicis. Note tendon slip (apl) from abductor pollicis longus into origin of abductor pollicis brevis. APB = abductor pollicis brevis; APL = abductor pollicis longus; FPB = flexor pollicis brevis; OP = opponens pollicis.

is unable to abduct the thumb away from the palm, it alone cannot produce true opposition.*

Adductor Pollicis

The adductor pollicis is the largest intrinsic muscle of the hand. It has two heads (Fig. 16). The transverse head is the more superficial and arises from the volar surface of the third metacarpal. The deeper oblique head arises from the ligaments covering the capitate and the trapezoid, as well as the distal part of the tunnel of the flexor carpi radialis. The two heads merge to form a short tendon which inserts into the ulnar sesmoid of the thumb. The sesmoid is embedded in the

* This should be remembered when testing median nerve function. Although the ulnar-innervated deep head of the flexor pollicis brevis may flex the metacarpal and the metacarpophalangeal joint, and thereby bring the tip of the thumb to the side of the index finger, this does not constitute true opposition. True opposition includes abduction of the first metacarpal away from the palm and rotation of the thumb and its metacarpal on its long axis. This combination of abduction and rotation makes possible the formation of a nearly round letter O by the thumb and index.

palmar ligament of the metacarpophalangeal joint of the thumb. Because the palmar ligament is firmly attached to the proximal phalanx and is loosely attached to the metacarpal, the adductor pollicis acts primarily on the proximal phalanx and only indirectly on the metacarpal. Nevertheless, the adductor pollicis is a powerful adductor of the first metacarpal as well as a flexor of the metacarpophalangeal joint of the thumb. The adductor pollicis has two secondary insertions. One is into the ulnar lateral tubercle of the proximal phalanx of the thumb. The other is into the extensor mechanism of the thumb (Fig. 81); therefore the adductor pollicis is an extensor of the distal joint of the thumb. The adductor pollicis is of paramount importance for pinch between thumb and fingertip. It is almost as important for grasp of large objects. The nerve supply is from the ulnar nerve.

HYPOTHENAR MUSCLES

The hypothenar muscles are the abductor digiti minimi, the flexor digiti minimi and the opponens

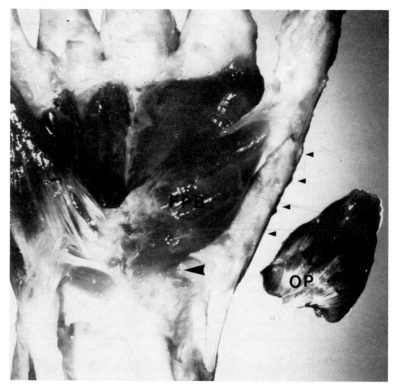

Figure 22. Opponens pollicis, detached from tendinous origin on flexor retinaculum and tubercle of the trapezium (large arrow). Note that insertion (small arrows) is proximal to metacarpophalangeal joint. FPB = flexor pollicis brevis; OP = opponens pollicis.

digiti minimi. They are covered by the thin hypothenar fascia which is continuous with the palmar aponeurosis.

Abductor Digiti Minimi

The abductor digiti minimi is the most superficial of the three hypothenar muscles (Fig. 24). Arising primarily from the pisiform and from the pisohamate ligament, it may have one or two bellies. The muscle has a dual insertion, one into the tubercle of the ulnar side of the base of the proximal phalanx, the other into the extensor mechanism. This muscle abducts the little finger and extends both interphalangeal joints. It is supplied by the ulnar nerve.

Flexor Digiti Minimi

The flexor digiti minimi brevis lies deep to the abductor digiti minimi (Fig. 25). Arising from the flexor retinaculum, the hook of the hamate and the pisohamate ligament, it inserts into the base of the proximal phalanx volar to the phalangeal insertion of the abductor. This muscle is a strong flexor of the metacarpophalangeal joint of the little finger. It also flexes and adducts the fifth

metacarpal, aiding in cupping the hand and in opposing the little finger to the thumb. The flexor brevis is innervated by the ulnar nerve.

Opponens Digiti Minimi

The opponens digiti minimi, also innervated by the ulnar nerve, is the smallest and the most deeply situated of the hypothenar group (Fig. 26). It arises from the hook of the hamate and from the pisohamate ligament and inserts along the length of the fifth metacarpal on the ulnar palmar border. The opponens digiti minimi flexes and adducts the fifth metacarpal toward the thumb.

The hypothenar muscles are variable. Any two of these muscles may be found fused. Either the abductor digiti minimi or the flexor digiti minimi may be absent. An accessory abductor digiti minimi may occasionally arise from the palmaris longus or from a wrist flexor in the distal forearm and extend obliquely over the abductor digiti minimi (Fig. 27).

INTEROSSEOUS MUSCLES

There are four dorsal interosseous muscles and three palmar interosseous muscles (Fig. 28). All

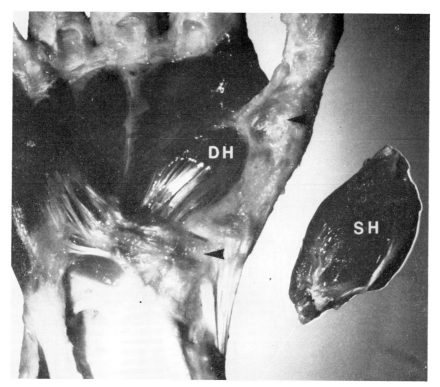

Figure 23. Flexor pollicis brevis, superficial head has been removed from origin on flexor retinaculum and tubercle of trapezium (lower arrow) and from insertion into radial sesmoid of metacarpophalangeal joint (upper arrow). DH = deep head of flexor pollicis brevis; SH = superficial head of flexor pollicis brevis.

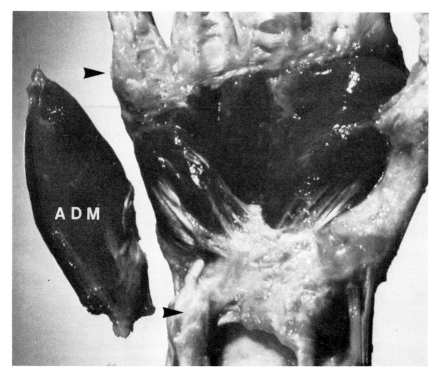

Figure 24. Abductor digiti minimi, detached from origin on pisiform and pisohamate ligament (lower arrow) and from insertion (upper arrow). ADM = abductor digiti minimi.

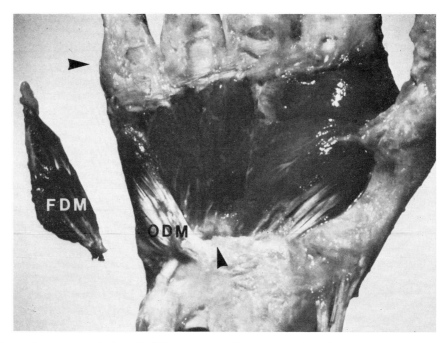

Figure 25. Flexor digiti minimi, detached from origin (lower arrow) and from insertion on proximal phalanx of little finger (upper arrow); FDM = flexor of digiti minimi; ODM = opponens digiti minimi.

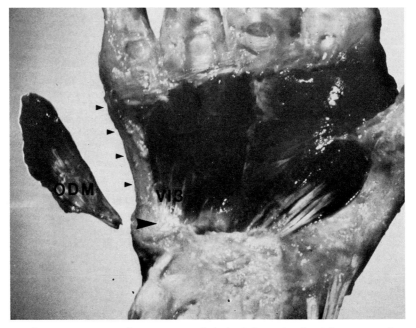

Figure 26. Opponens digiti minimi, tendinous origin detached from hook of hamate and pisohamate ligament (large arrow). Note muscular rather than tendinous insertion along length of fifth metacarpal shaft (small arrows). ODM = opponens digiti minimi; V13 = third volar interosseous.

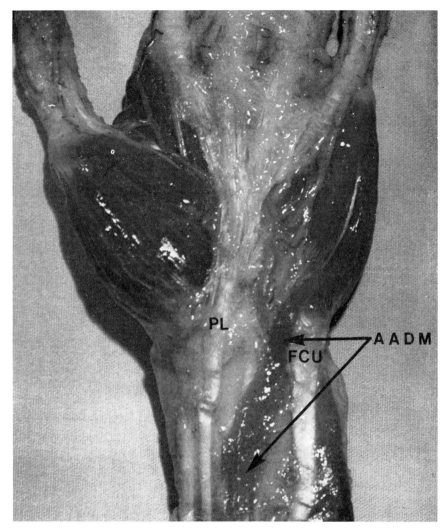

Figure 27. Accessory abductor digiti minimi. AADM = accessory abductor digiti minimi; FCU = flexor carpi ulnaris; PL = palmaris longus.

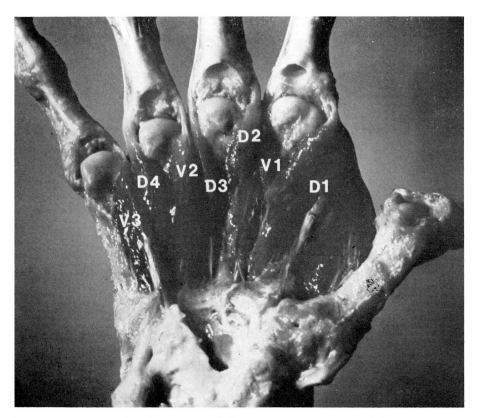

Figure 28. The interosseous muscles. Natatory and deep transverse metacarpal ligaments, as well as palmar ligaments of each metacarpophalangeal joint removed. D1 = first dorsal interosseous; D2 = second dorsal interosseous, etc.; VI = first volar interosseous, etc.

of the interossei are supplied by the ulnar nerve. Each dorsal interosseous has two muscle bellies which arise from the two adjacent metacarpals. The dorsal interossei insert distal to the metacarpophalangeal joint. The first inserts on the radial side of the index and is therefore an abductor of the index. The second inserts on the radial side of the long finger while the third inserts on the ulnar side of the same finger. Considering the third ray to be the central axis of the hand, the second dorsal interosseous is then a radial abductor of the long finger while the third is an ulnar abductor of the same finger. The fourth dorsal interosseous, inserting on the ulnar side of the ring finger, is the abductor of that finger.

The palmar interossei, each with a single belly, arise from the more palmar portion of the metacarpals. The first palmar interosseous arises from the second metacarpal, the second from the fourth metacarpal, and the third from the fifth metacarpal. They insert respectively on the ulnar side of the index finger and on the radial sides of the ring and little fingers. Therefore, the first palmar inter-

osseous is an adductor of the index, the second adducts the ring finger, and the third is an adductor of the little finger. The tendons of both the palmar and the dorsal interossei pass distally onto the fingers dorsal to the deep transverse metacarpal ligament (Fig. 29) of the palm. The lumbricals pass volar to this ligament.

The plane of the interosseous tendons is volar to the axis of flexion of the metacarpophalangeal joints. Therefore, the interossei are flexors of those joints. Because of their contributions to the extensor mechanism, they also extend the interphalangeal joints (Ch. VIII).

METACARPALS

The second, third, fourth and fifth metacarpals constitute the bulk of the bony framework of the palm. At their bases, the metacarpals are in close apposition and are tightly tethered together. The third metacarpal may be thought of as rigid center post. The second metacarpal is slightly mobile in relationship to the third metacarpal. The fourth metacarpal is more mobile than the second, and

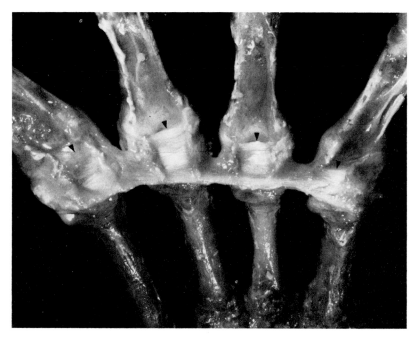

Figure 29. Deep transverse metacarpal ligament and the metacarpals. Palmar ligament of each metacarpophalangeal joint has been preserved. Note that the deep transverse metacarpal ligament connects the palmar ligament (arrows) of the metacarpophalangeal joint together.

the fifth is the most mobile of the four. The metacarpals are more loosely tethered at their heads than at their bases. Distally, the metacarpals are indirectly connected by the deep transverse metacarpal ligament (Fig. 29). This ligament is an interconnection between the palmar ligaments of the metacarpophalangeal joints. There is ample room for the interosseous muscles between the metacarpals, as the bases and the heads of the metacarpals are wider than their shafts.

CHAPTER III

THE THUMB

Skin and Subcutaneous Tissue

THE THICKNESS OF THE SKIN and of the subcutaneous layer of fat of the thumb varies according to location. The dorsal skin and subcutaneous fat layer are thinnest. The skin and the layer of fat over the volar surface of the proximal phalanx of the thumb are of moderate thickness and similar to the volar skin of the fingers (Ch. IV). The skin and the underlying layer of fat over the volar surface of the distal phalanx is much thicker. The distal pulp space of the thumb extends from the tip of the thumb proximally over the volar surface of the distal phalanx, almost to the interphalangeal joint. The pulp space is sealed proximally by strong, inelastic ligaments which fasten the skin to the distal phalanx and to the insertion of the flexor pollicis longus. The space is crisscrossed by fibrous septa which are perpendicular to the skin and which minimize skin mobility.*

Proper Palmar Digital Arteries

The proper palmar digital arteries of the thumb lie beneath the subcutaneous fat. The two arteries arise from the bifurcation of the princeps pollicis artery, which is a branch of the radial artery. The arteries course distally, one on each side of the tendon of the flexor pollicis longus. The arteries arborize in the distal pulp space, and terminal branches anastamose over the distal half of the volar surface of the distal phalanx, forming the counterpart of the anastamoses in the fingers (Fig. 17).

Proper Palmar Digital Nerves

The two proper palmar digital nerves of the thumb arise in the palm from the median nerve (Ch. II). In the proximal palm, both nerves lie to the ulnar side of the tendon of the flexor pollicis longus. The proper digital nerve to the radial side, however, crosses to the radial side of the thumb at a point proximal to the metacarpophalangeal joint (Fig. 30). While crossing the tendon the nerve is superficial to it.

Distally along the proximal phalanx, the two digital nerves run parallel courses, closely juxtaposed to either side of the flexor pollicis longus. The proper digital nerves arborize over the volar surface of the distal phalanx, forming a rich plexus in the distal pulp.

Flexor Pollicis Longus

The tendon of the flexor pollicis longus in the proximal palm is covered by the muscle bellies of the abductor pollicis brevis and of the opponens pollicis. As the tendon approaches the metacarpophalangeal joint and becomes subcutaneous (Fig. 31), its fibrous sheath becomes thickened over the metacarpophalangeal joint and the proximal portion of the proximal phalanx. The tendon crosses the metacarpophalangeal joint in a shallow trough formed by the palmar ligament buttressed on either side by a sesmoid bone. The flexor pollicis longus continues distally across the interphalangeal joint. It is supplied by the median nerve.

PINCH AND GRASP

The two most important functions of the hand are pinch and grasp. The thumb plays an essential role in the performance of both. The bony architecture of the thumb and the arrangements of its eight muscles and of the ligaments of its joints are adapted to these roles.

Pinch requires the tip of the thumb to either

---

* The relative lack of skin mobility in this area is important for precise pinch. If the skin were mobile, small objects would slip from between the thumb and the fingertips.

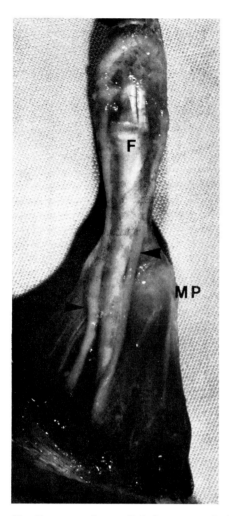

Figure 30. Proper palmar digital nerves of thumb. Arrows = proper digital nerves; F = flexor pollicis longus; MP = metacarpophalangeal joint.

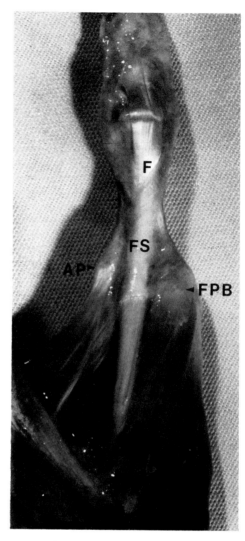

Figure 31. Flexor pollicis longus. Filmy synovial tendon sheath has been removed to demontrate thick fibrous sheath better. AP = adductor pollicis insertion into ulnar sesmoid; F = flexor pollicis longus; FPB = flexor pollicis brevis insertion into radial sesmoid; FS = fibrous sheath of flexor pollicis longus.

oppose the tip or the side of the index finger alone or to oppose simultaneously the tips of both the index and long fingers to form a three-jawed chuck. Pinch emphasizes precision. In contrast, when large objects are grasped power is often needed. In these instances the thumb is called upon to match the strength of the four fingers working together.

The functions of pinch and grasp require the thumb to be in a position of opposition to one or more fingers. If the thumb were fixed permanently in this position, it would be in the way for the more primitive activities of the hand, such as

pushing and patting. The mobility of the first metacarpal on the trapezium allows the thumb to be brought into the position of opposition when necessary for pinch and for grasp and also be brought back to the plane of the fingers to perform other functions of the hand.

BONES AND JOINTS

The design of the bones and joints of the thumb reflect the requirements for both strength and mobility imposed by the roles of the thumb in pinch and grasp. The thumb is stout compared to the fingers. The metacarpal and the proximal phalanx are broad and short. There is no middle phalanx (Figs. 32 and 33). The complex contours of the surface of the trapezium which articulates with the first metacarpal makes possible the synchronous movements of extension, abduction, rotation, flexion and adduction of the first metacarpal, which constitutes opposition of the thumb. The virtual immobility of the intercarpal joints during opposition is demonstrated in Figures 34 and 35.*

Metacarpophalangeal and Interphalangeal Joints

The metacarpophalangeal and the interphalangeal joints of the thumb are diarthrodial joints. They both have collateral ligaments on each side which provide lateral stability. The collateral ligaments of the metacarpophalangeal joint of the

* Although only motion between the first metacarpal and the trapezium takes place during opposition, loss of motion at this joint by arthrodesis or by arthritis does not necessarily eliminate opposition. The adjacent intercarpal joints may compensate to varying degrees.

thumb are stout and are taut both in flexion and in extension (Fig. 36). Therefore, there is little lateral play at this joint. For practical purposes, there is no adduction or rotation at the metacarpophalangeal joint of the thumb. The arc of flexion and extension of this joint is perhaps two-thirds that of the metacarpophalangeal joints of the fingers.

The interphalangeal joint of the thumb is a diarthrodial ginglymus joint similar to the distal interphalangeal joints of the fingers. It possesses a range of flexion and extension comparable to the terminal finger joints, and it permits no appreciable motion in any other plane.

Palmar Ligaments

The metacarpophalangeal joint and the interphalangeal joint of the thumb each have a palmar ligament which lies between the volar surface of the joint and the tendon of the flexor pollicis longus. Two sesmoids are embedded in the palmar ligament of the metacarpophalangeal joint. The

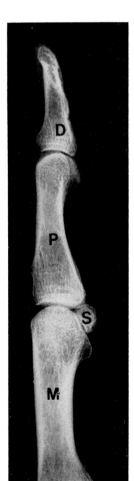

Figure 32. X-ray of thumb, lateral view. D = distal phalanx; M = metacarpal; F = proximal phalanx; S = sesmoid.

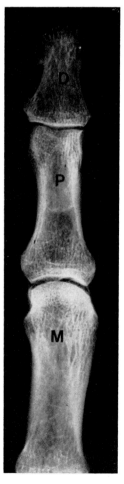

Figure 33. X-ray of thumb, anteroposterior view. D = distal phalanx; M = metacarpal; P = proximal phalanx.

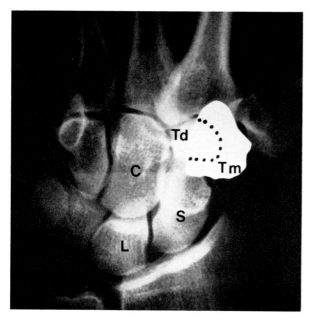

Figure 34. Anteroposterior X-ray of hand, thumb unopposed (in same plane as fingers). C = capitate; L = lunate; M = first metacarpal; S = scaphoid; Td = trapezoid; Tm = trapezium.

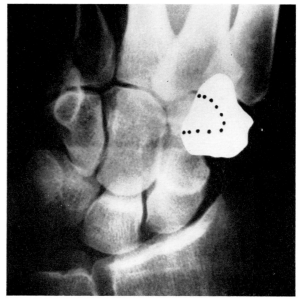

Figure 35. Anteroposterior X-ray of the hand, thumb in maximum opposition. Note that virtually all motion has taken place at trapeziometacarpal joint.

palmar ligament is thickest at its distal end and is firmly attached to the proximal end of the proximal phalanx. The proximal end of the palmar ligament is thinner and is attached loosely to the metacarpal. The palmar ligament of the interphalangeal joint is like that of the metacarpophalangeal joint, but it usually contains no sesmoids.

MUSCLES OF THE THUMB

The thumb is positioned and stabilized by four extrinsic muscles and by four intrinsic thenar muscles. The extrinsic muscles of the thumb are the flexor pollicis longus (described above) and the extensors pollicis longus and brevis and the abductor pollicis longus (Ch. VIII). The intrinsic muscles are the abductor pollicis brevis, the opponens pollicis, the flexor pollicis brevis, and the adductor pollicis.

Thenar Muscles

The thenar muscles differ in several important respects from the intrinsic muscles of the fingers. The intrinsic muscles of the thumb as a group are substantially larger than those of any finger. Further, the origins and the insertions of the thenar muscles are not comparable to their counterparts of the fingers. Therefore, the actions of the thenar muscles are different from the actions of the inter-

osseous muscles and of the lumbricals. The abductor pollicis brevis, the opponens pollicis, the flexor pollicis brevis, and the adductor pollicis all arise from the carpal bones and their associated ligaments (Ch. II). Since these muscles span the trapeziometacarpal joint, they act on that joint and abduct, rotate, and flex and adduct the first metacarpal.

The opponens pollicis inserts broadly along the shaft of the first metacarpal and therefore has no effect on the distal joints. The other three thenar muscles have insertions distal to the metacarpophalangeal joint and act on the joint and on the interphalangeal joint as well. The adductor pollicis and the flexor pollicis brevis insert primarily into the ulnar and radial sesmoids, respectively. The sesmoids of the metacarpophalangeal joint of the thumb are embedded in the palmar ligament which is attached principally to the proximal phalanx. Expansions of these muscles provide secondary insertions into the extensor mechanism of the thumb (Ch. VII). Therefore, these two muscles are strong flexors of the metacarpophalangeal joint but are only weak extensors of the terminal joint.

The abductor pollicis brevis inserts by a tendon into the lateral capsule of the metacarpophalangeal joint (Fig. 37) and has a secondary insertion into the extensor mechanism. The only action of the abductor pollicis brevis on the metacarpophalangeal joint is flexion, as that joint permits no significant amount of abduction or rotation. However, the abductor pollicis brevis is an effective abductor and rotator, as well as a flexor, of the first metacarpal. In addition, this muscle is an extensor of the interphalangeal joint of the thumb.

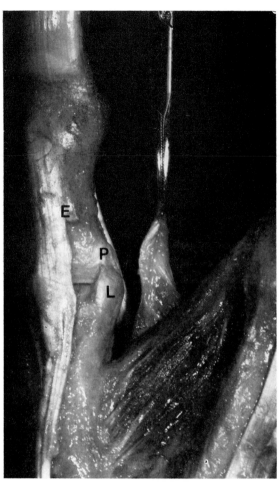

Figure 36. Thumb metacarpophalangeal joint, ulnar aspect. The adductor pollicis tendon (in clamp) has been detached from ulnar sesmoid (not visible), ulnar lateral tubercle of the proximal phalanx (P), and from extensor mechanism (E); joint capsule excised, exposing collateral ligament (L). Note that collateral ligament is taut, even though joint is in extension.

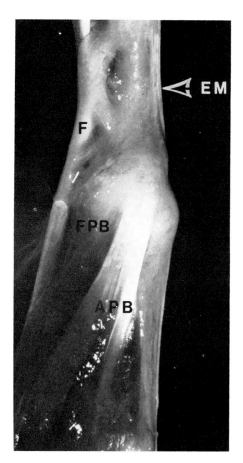

Figure 37. Thumb metacarpophalangeal joint, radial aspect (enlarged 2×). APB = abductor pollicis brevis; EM = extensor mechanism; F = flexor pollicis longus. Note some fibers from flexor pollicis brevis extend into extensor mechanism.

CHAPTER IV

THE FINGERS

THE VOLAR ASPECT OF THE FINGERS is covered by highly specialized skin. It has a thick cornified layer, is quite inelastic, and is well endowed with sensory nerve endings as well as sweat glands. Papillary ridge traverse the surface.[2]

Each finger has three flexion ridges (Fig. 7). The palmar digital creases lie at the level of the midportion of the shaft of the proximal phalanx. The interdigital folds lie in the same plane as the palmar digital creases. The second and third flexion creases lie over the proximal and distal interphalangeal joints. The flexion creases divide the subcutaneous fat of the fingers into discrete fat pads overlying each phalanx.

Cutaneous Ligaments

The volar skin of the fingers is relatively immobile. In addition to its intrinsic thickness and inelasticity, it is tethered to the underlying structures by multiple ligaments.[3]

From the natatory ligament and the palmar aponeurosis longitudinal fibers pass distally along the volar-lateral aspect of the finger (Fig. 38). The fibers blend into the dermis distal to the proximal interphalangeal joint.* Short, fibrous septa anchor the fat pads to the flexor tendon sheath.

On each lateral side of the finger, two ligamentous complexes fasten the skin to the phalanges and to the flexor tendon sheath, forming an incomplete tunnel for each neurovascular bundle.

[2] Michael Hartz, "The Dermal Papillae in the Fingertip," *Plastic and Reconstructive Surgery* (Baltimore), Vol. 45 (1970), p. 141.

[3] Lee W. Milford, Jr., *Retaining Ligaments of the Digits of the Hand. Gross and Microscopic Anatomic Study* (Philadelphia, W. B. Saunders, 1968).

* These distal extensions of the pretendinous bands may be the fibers which are often hypertrophied in cases of Dupuytren's contractures, producing flexion deformities of the proximal interphalangeal joint.

The two complexes are named Grayson's and Clelland's ligaments.

Grayson's ligament is a delicate sheet of transverse, essentially parallel fibers which fasten the skin to the flexor tendon sheath (Fig. 39). The sheet of fibers extend from the middle of the proximal phalanx to the distal interphalangeal joint. This ligamentous complex lies superficial to the neurovascular bundle and is the more volar of the two complexes.

Clelland's ligament is comprised of two pairs of fibrous bands, one pair at the proximal interphalangeal joint and one pair at the distal interphalangeal joint (Fig. 40). The proximalmost band attaches the skin to the distal end of the proximal phalanx. The second band attaches the skin to the proximal end of the middle phalanx. The third band connects the distal end of the middle phalanx to the skin, and the fourth connects the proximal end of the distal phalanx to the skin. The bands of Clellands' ligament pass deep to the neurovascular bundle.

Distal Pulp Space

The distal pulp space of the volar aspect of the finger extends from the tip of the finger proximally over the distal phalanx, almost to the distal interphalangeal flexion crease (Fig. 41). The pulp space is sealed proximally by strong, inelastic ligaments which fasten the skin to the distal phalanx and to the insertion of the profundus tendon. The space is crisscrossed by fibrous septa which are perpendicular to the skin and which minimize skin mobility.† The nail plate stabilizes the pulp at the distal tip and the lateral margins. The fat-laden pulp space contains an extensive vascular network

† If the skin in this area were mobile, precise pinch would be difficult if not impossible. Small objects would slip from between the thumb and fingers.

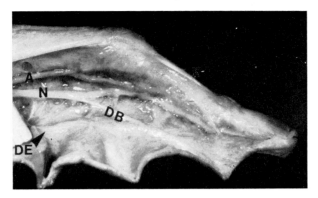

Figure 38. Distal extensions of the pretendinous bands. Skin incised longitudinally along dorsum of finger and reflected volarward; dorsalmost skin is at bottom of illustration. A = proper palmar digital artery; DB = dorsal branch of proper palmar digital nerve; DE = distal extension of pretendinous band; N = proper palmar digital artery.

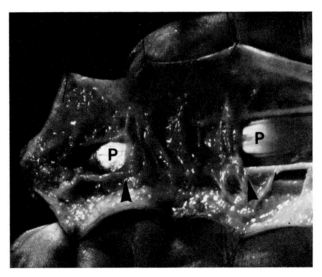

Figure 39. Grayson's ligaments. Volar aspect of finger; skin incised longitudinally and reflected to either side. Large arrows = proximal and distal borders of Grayson's ligament which connects skin to tendon sheath; small arrows = proper palmar digital nerves, which pass deep to Grayson's ligament; P = flexor digitorum profundus.

as well as terminal branches of the digital nerves (Fig. 17).

Digital Arteries

The origins of the volar arterial supply of the digits vary widely.[4] Figure 13 illustrates a common arrangement. The proper palmar digital artery to the ulnar side of the little finger is a direct branch of the ulnar artery. The superficial palmar arch gives off three common palmar digital arteries.

[4] Sherman S. Coleman and Barry J. Anson, "Arterial Patterns in the Hand Based Upon a Study of 650 Specimens," *Surgery, Gynecology and Obstetrics* (Chicago), Vol. 113 (1961), p. 409.

They extend distally to the second, third and fourth interdigital folds; bifurcate; and form proper digital arteries. These arteries pass between the natatory ligament and the lumbrical muscles to enter the fingers. The proper digital artery to the radial side of the index finger is a direct branch of the superficial palmar arch.

The proper palmar digital arteries lie dorsal to the proper digital nerves in their course over the proximal and middle phalanges. Beginning at the proximal interphalangeal joint area the proper palmar digital arteries give off multiple dorsal branches (Fig. 83). (See Ch. VIII.)

On each side of each finger the proper palmar digital nerve and the proper palmar digital artery

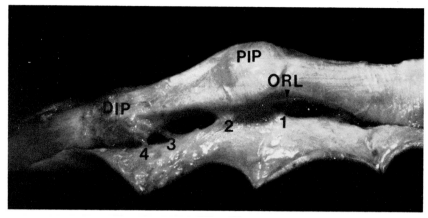

Figure 40. Clelland's ligament. Dorsolateral view of finger, skin incised longitudinally along dorsum and retracted volarward on either side. DIP = distal phalangeal joint; ORL = oblique retinacular ligament; PIP = proximal interphalangeal joint; 1, 2, 3, 4 = the four bands of Clelland's ligament.

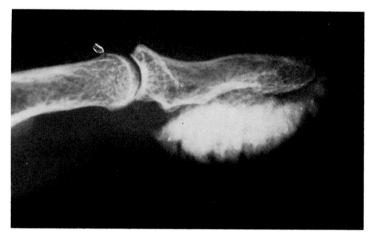

Figure 41. Distal pulp space, injected with radio-opaque material. Note subcutaneous indentations into radio-opaque material produced by fibrous septa which connect skin to bone.

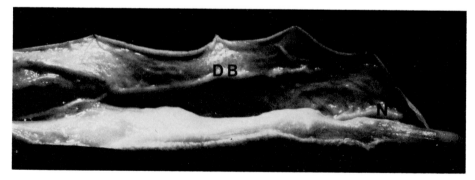

Figure 42. Dorsal branch of proper palmar digital nerve. Dorsal view of finger, skin incised longitudinally and retracted volarward on one side. DB = dorsal branch; N = main trunk of proper palmar digital nerve just before arborization.

lie in close juxtaposition. Enveloped in a thin fascial covering, the nerve and artery comprise a neurovascular bundle. The neurovascular bundles to the adjacent surfaces of the index, long, ring and little fingers pass into the fingers dorsal to the natatory ligament. The neurovascular bundles to the radial side of the index finger and to the ulnar side of the little finger course obliquely across the metacarpophalangeal joints through the subcutaneous fat. The neurovascular bundles extend distally on the volar-lateral aspects of the fingers. They pass through the incomplete tunnels formed by Clelland's and Grayson's ligaments (Figs. 39 and 40) and are separated from the flexor tendon sheaths by areolar fat.

The proper palmar digital nerve gives off a major dorsal branch at the level of the proximal phalanx (Figs. 38, 42). The main trunk continues distally to the distal phalanx where it arborizes to form a rich sensory plexus supplying the skin over the pulp of the fingers.

Flexor Tendon Sheaths

The tendons of the flexor digitorum profundus and of the flexor digitorum superficialis of each finger are covered by a thin synovial sheath before reaching the metacarpophalangeal joint. At this joint, the tendons acquire a tough outer fibrous sheath. The volar and lateral walls of the outer sheath are formed by firm fibrous tissue. The dorsal wall is formed by the periosteum covering the phalanges and by the palmar ligaments which bridge the joints. This complex is often called the fibro-osseous tunnel.

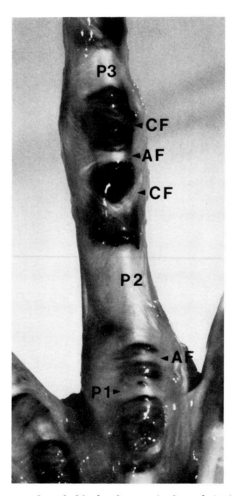

Figure 43. Flexor tendon sheath injected with black silicone (enlarged 2×). Where well developed, outer fibrous sheath is opaque. Where opaque outer fibrous sheath is sparsely developed, the black contrast material can be seen through semitransparent inner synovial sheath. AF = annular fibers; CF = cruciate fibers; P1 = pulley 1; P2 = pulley 2; P3 = pulley 3.

Pulleys

The term *pulley* is used to designate a section of the fibrous sheath which holds the flexor tendons in close proximity to the underlying bone. In those sections of the fibrous sheath designated as pulleys, the lateral and volar walls of the sheath consist of firm transverse bands of fibrous tissue. Between the pulleys the walls of the fibrous sheath are composed of cruciate and annular fibers.

The most proximal pulley lies over the palmar ligament of the metacarpophalangeal joint and is quite narrow (Fig. 43). The second and third pulleys are quite broad and insert respectively on the volar-lateral ridges of the proximal and middle phalanges. The fourth pulley, at the base of the distal phalanx, is less discrete.

Flexor Digitorum Superficialis

Midway through the second pulley the tendon of the flexor digitorum superficialis splits in the coronal plane into two slips (Figs. 44 and 45). Each slip of the tendon rotates outward through an arc of 180 degrees around the tendon of the flexor digitorum profundus. As the slips of the flexor digitorum superficialis gain new positions dorsal to the profundus, each slip again divides in two (Fig. 46). The fibers of the lateral half of each slip continue in a straight distal course. The medial fibers of each slip course obliquely across the midline of the finger forming the chiasm. Continuing distally across the proximal interphalangeal joint, the medial fibers of the radial tendon slip merge distally with the lateral fibers of the

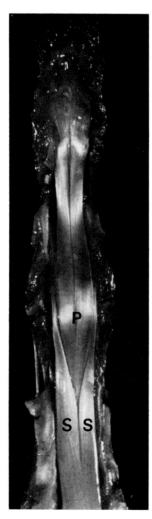

Figure 44. Flexor tendon sheaths. P = flexor digitorum profundus; S = slips of flexor digitorum superficialis.

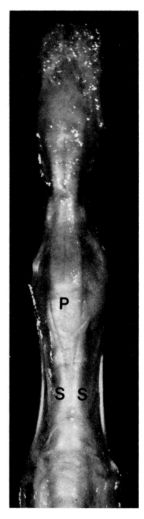

Figure 45. Flexor tendons, sheaths removed. P = flexor digitorum profundus; S = slips of flexor digitorum superficialis.

ulnar tendon slip. Conversely the medial fibers of the ulnar tendon slip merge distally with the lateral fibers of the radial tendon slip. Each merged slip then inserts into the volar-lateral aspect of the middle phalanx.

The flexor digitorum superficialis is primarily a flexor of the proximal interphalangeal joint. The superficialis of the little finger is often small and ineffective. The four flexors digitorum superficialis are all supplied by the median nerve.

Flexor Digitorum Profundus

The flexor digitorum profundus inserts distally into the base of the distal phalanx (Fig. 45). This muscle is an effective flexor of both of the interphalangeal joints. The muscle belly of the profundus of the index finger is normally separate

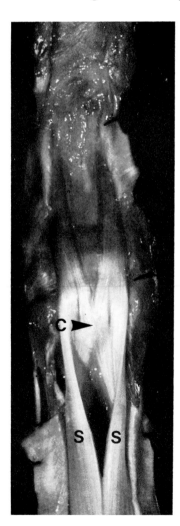

Figure 46. Flexor digitorum superficialis (enlarged 2×). Profundus removed. C = chiasm; S = slips of flexor digitorum superficialis.

from the bellies of the other profundi. However, the muscle bellies of the flexors digitorum profundus of the other three fingers are intimately attached to one another. Therefore, the profundi of the ulnar three digits are incapable of completely independent action. The flexors digitorum profundus to the index and long fingers are innervated by the median nerve, while those to the ring and little fingers are supplied by the ulnar nerve.

Vincula

The vincula (Fig. 47) are strands of mesotenon which provide blood supply to the flexor tendons as they course through the fibro-osseous tunnel. Each intrinsic flexor tendon has a long and a short vinculum. The long vinculum to the flexor digitorum superficialis arises from the proximal portion of the proximal phalanx. Twin strands of the vinculum course to each of the two tendon slips. The short vinculum extends from the distal portion of the proximal phalanx to the decussation of the flexor digitorum superficialis.

The long vinculum of the flexor digitorum profundus passes to this tendon from the decussation. The short vinculum has a broad base which extends from the insertion of the flexor digitorum superficialis to the palmar ligament of the distal interphalangeal joint. The viniculum is attached to the distal end of the tendon of the flexor digitorum profundus.

PHALANGES

The phalanges are the bony framework of the fingers. Each finger possesses three such bones, termed the proximal, middle and distal phalanges (Figs. 48 and 49).

Proximal Phalanx

The base of the proximal phalanx is wider than the shaft. The articular surface is concave in both the anteroposterior and the lateral planes. The proximal phalanx has a longitudinal ridge on each side of its volar surface which extends the length of the shaft. These ridges are continuous except for a short section at the junction of the middle and distal thirds. The longitudinal trough formed by the two volar ridges accommodates the flexor tendons. The dorsum of the shaft is smooth and is mildly convex both longitudinally and transversely. The head of the phalanx is composed of two condyles and a shallow intercondylar notch.

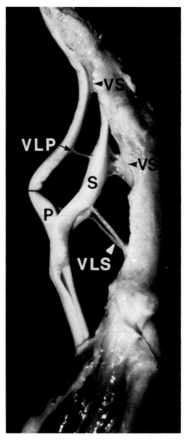

Figure 47. Vincula (lateral view of finger). P = flexor digitorum profundus; S = flexor digitorum superficialis; VLS = double long vinculum to superficialis; VLP = long vinculum to profundus; VS = short vinculum.

Middle Phalanx

The middle phalanx is shorter than the proximal phalanx. The base is wider than the shaft. Twin concavities on the articular surface of the proximal end of the phalanx receive the condyles of the head of the proximal phalanx. The longitudinal concavity of the volar aspect of the phalanx is limited to the midportion of the shaft. The dorsum of the shaft is smooth and convex. The head of the middle phalanx is similar to that of the proximal phalanx, possessing two condyles and a shallow condylar notch.

Distal Phalanx

The distal phalanges of the four fingers are usually of nearly equal length. The base of the distal phalanx is as wide as the head of the middle phalanx. The articular surface features twin concavities which correspond to the condyles of the head of the middle phalanx. The shaft of the distal phalanx, narrower than the shaft of the middle phalanx, tapers and then flares into a terminal tuft which provides broadened support for the overlying soft tissues of the fingertip.

JOINTS

The finger joints are all diarthrodial. The metacarpophalangeal joints are condyloid, and the interphalangeal are ginglymus. Strong collateral ligaments provide lateral stability to each joint. The collateral ligaments are thickest and strongest at the metacarpophalangeal joints (Fig. 50) and are least well developed at the distal interphalangeal joints (Fig. 53).*

* These structural variations may reflect the difference between the length of the lever which applies stresses to the collateral ligaments of a metacarpophalangeal joint as compared to the length of the levers which act on the more distal joints. The collateral ligaments of the metacarpophalangeal joints are subject to stresses which are amplified by the long lever of the full length of the finger. In contrast, at a distal interphalangeal joint, the short distal phalanx is the lever. Even at a proximal interphalangeal joint, the lever formed by the middle and the distal phalanges may be only a little more than half the length of the entire finger.

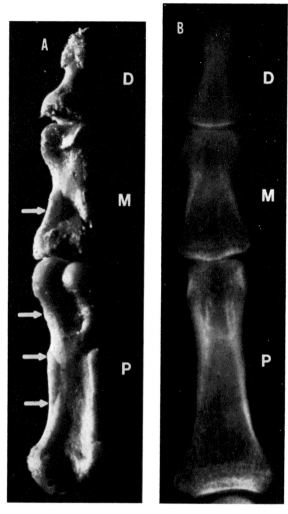

Figure 48. The phalanges (enlarged 2×). A. Volar view of denuded phalanges to show lateral volar ridges. B. Anteroposterior X-ray. Arrows = lateral volar ridges; D = distal phalanx; M = middle phalanx; P = proximal phalanx.

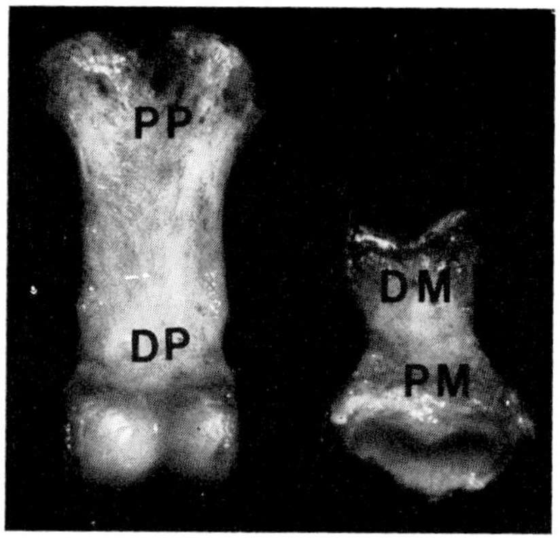

Figure 49. Proximal interphalangeal articular surfaces (enlarged 1.5×). DM = distal end middle phalanx; DP = distal end proximal phalanx; PM = proximal end middle phalanx; PP = proximal end proximal phalanx.

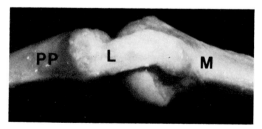

Figure 50. Metacarpophalangeal collateral ligament, lateral view. Note that metacarpal attachment is closer to dorsum (upper) and phalangeal attachment is more volar. L = ligament; M = metacarpal; PP = proximal phalanx.

Palmar Ligaments

At each metacarpophalangeal joint and at each interphalangeal joint a palmar ligament covers the articular surface beneath the flexor tendons (Figs. 51, 52 and 53). The palmar ligaments are fibrocartilaginous plates and are much thicker at the distal end than at the proximal. The distal attachments are strong whereas the proximal attachments are membranous and tenuous.

Metacarpophalangeal Joints

The mechanics of the metacarpophalangeal joint are largely determined by the shape of the metacarpal head and by the placement of the collateral ligaments. The dorsal half of the metacarpal head is essentially spherical in shape whereas the palmar surface is bicondylar (Fig. 54). When the metacarpal head is viewed on end it is seen to be wider on the palmar aspect than on the dorsal aspect (Fig. 55). The attachment of the collateral

ligament on the metacarpal head is eccentric. The ligament is attached at a point which is both dorsal to and distal to the axis of flexion of the joint (Fig. 50). The combination of the eccentric attachment and the greater transverse width of the metacarpal head on its palmar aspect account for the fact that the collateral ligament is slack when the joint is extended and taut when the joint is flexed.* Flexion of the joint increases the straight-line distance between the metacarpal attachment and the phalangeal attachment of the collateral

---

* The laxity of the collateral ligaments when the metacarpophalangeal joints are extended must be borne in mind whenever it is necessary to immobilize one or more of these joints. When the joints are immobilized in extension, the collateral ligaments may shorten over time, particularly in older people. If this occurs, the joints cannot be flexed when immobilization is discontinued. Therefore, the metacarpophalangeal joints of the fingers should be immobilized in flexion except when there is a compelling reason to hold them in extension.

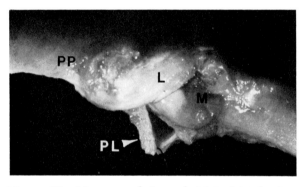

Figure 51. Metacarpophalangeal joint, lateral view. Palmar ligament split longitudinally and lateral portion removed to show its distal thickness and its proximal tenuousness. L = collateral ligament; M = metacarpal; PP = proximal phalanx; PL = palmar ligament.

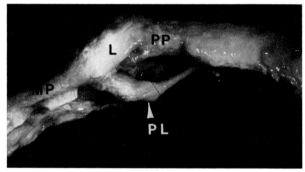

Figure 52. Proximal interphalangeal joint, lateral view. L = collateral ligament; MP = middle phalanx; PL = palmar ligament; PP = proximal phalanx.

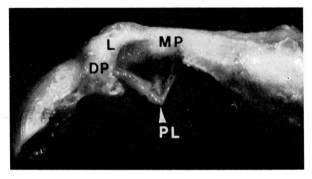

Figure 53. Distal interphalangeal joint, lateral view. DP = distal phalanx; L = collateral ligament; MP = middle phalanx; PL = palmar ligament.

ligament, thereby tightening the ligament. Further, when the joint is flexed, the collateral ligament is additionally stretched by the lateral bulge of the palmar portion of the metacarpal head. The ligament is under maximum tension when the joint is flexed approximately sixty-five degrees. The

joint is easily distracted when the joint is extended (Figs 56 and 57).

When the metacarpophalangeal joint is in extension, the combination of ligamentous laxity and the relatively spherical contour of the dorsal portion of the metacarpal head permit abduction

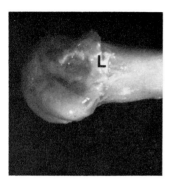

Figure 54. Metacarpal head, lateral view. Note lateral flaring of condyles on volar aspect of head beneath attachment of collateral ligament (L).

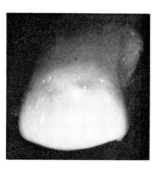

Figure 55. Metacarpal, viewed on end (enlarged 2×). Note that volar portion of head (lower) is broader than dorsal aspect.

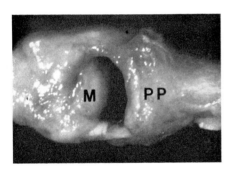

Figure 56. Metacarpophalangeal joint, dorsal view, distracted and extended (enlarged 2×). All structures except collateral ligaments removed. Note gap between metacarpal and proximal phalanx. M = metacarpal; PP = proximal phalanx.

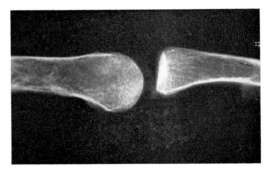

Figure 57. Metacarpophalangeal joint, dorsal view. X-ray of specimen shown in Figure 57.

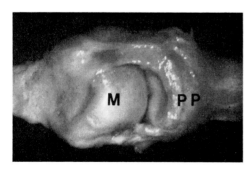

Figure 58. Metacarpophalangeal joint, dorsal view, distracted and flexed (enlarged 2×). Note metacarpal head is in firm contact with phalanx. M = metacarpal; PP = proximal phalanx.

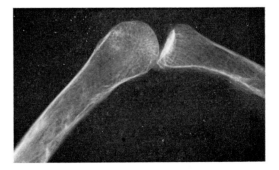

Figure 59. Metacarpophalangeal joint, dorsal view. X-ray of specimen shown in Figure 59.

and adduction of the joint, as well as flexion and extension. Conversely, when the joint is in flexion, the articular surfaces are pressed firmly together (Figs. 58 and 59). Furthermore, when the metacarpophalangeal joint is flexed, the proximal phalanx articulates with the broad palmar surface of the metacarpal head. The combination of a broad (rather than spherical) articular surface and of taut collateral ligaments prevent abduction and adduction when the joint is flexed.

Interphalangeal Joints

The heads of both the proximal and middle phalanges are bicondylar in shape. The condyles articulate with twin concavities in the bases of the middle and distal phalanges. The collateral ligaments of the interphalangeal joints are taut in both flexion and in extension. The combination of the articular configurations and the taut collateral ligaments limit the proximal and distal interphalangeal joints to the motions of flexion and extension.

CHAPTER V

RADIAL SIDE OF THE HAND

Anatomical Snuffbox

THE CENTRAL TOPOGRAPHIC FEATURE of the radial side of the hand is a triangular depression named the anatomical snuffbox. The base of the triangle is formed by the styloid process of the radius. The dorsal side of the triangle is formed by the tendon of the extensor pollicis longus. The tendons of the abductor pollicis longus and of the extensor pollicis brevis comprise the volar side of the triangle.

The borders of the snuffbox are brought into relief by simultaneously extending and abducting the thumb forcibly, while holding the wrist in neutral extension (Fig. 60). The scaphoid lies beneath the snuffbox (Figs. 63, 64 and 65).*

* The scaphoid is subject to fracture during a fall onto the outstretched hand. Initial X-ray examination may be negative. Tenderness in the snuffbox in such cases is

Radial Nerve

The sensory branch of the radial nerve passes over the radial styloid process in close proximity to the common sheath of the abductor pollicis longus and of the extensor pollicis brevis (Fig. 62).† The nerve traverses the radial side of the wrist deep to the subcutaneous fat and superficial to the tendons of the extensors pollicis, of the abductor pollicis longus, and of the extensors carpi radialis. As it courses over these tendons, this sensory branch of the radial nerve divides into smaller

strongly suggestive of occult fracture of the scaphoid. X-rays should be repeated in two or three weeks.

† The radial nerve is subject to injury here during surgical procedures for tenosynovitis of these tendon sheaths. The resulting sensory loss is rarely a significant problem, but the development of a painful neuroma may be exceedingly troublesome.

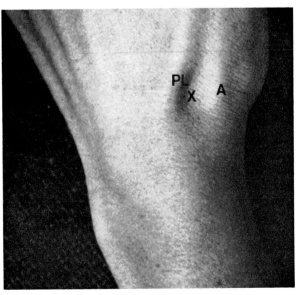

Figure 60. Anatomical snuffbox. A = combined sheath of abductor pollicis longus and extensor pollicis brevis; PL = extensor pollicis longus; X = snuffbox.

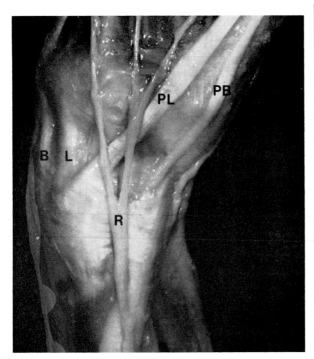

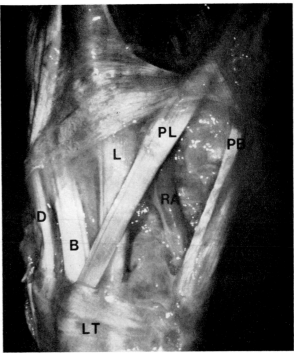

Figure 61. Radial nerve sensory branch. B = extensor carpi radialis brevis; L = extensor carpi radialis longus; PB = extensor pollicis brevis; PL = extensor pollicis longus; R = radial nerve.

Figure 62. Tendons and radial artery. A = abductor pollicis longus; B = extensor carpi radialis brevis; D = extensor digitorum; L = extensor carpi radialis longus; LT = Lister's tubercle; PB = extensor pollicis brevis; PL = extensor pollicis longus; RA = radial artery.

branches which supply the dorsum of the radial portion of the hand and the corresponding digits. The distribution of the terminal branches of the radial nerve is described in Chapters VII and VIII.

TENDONS

Extensor Pollicis Longus

The tendon of the extensor pollicis longus has a straight course in the distal forearm. It passes ulnar to Lister's tubercle and then angles toward the thumb (Figs. 62 and 71). This tubercle is situated in the center of the dorsum of the radius, just proximal to the radiocarpal joint (Fig. 65). It is easily palpated and is a constant landmark. The extensor pollicis longus tendon continues distally along the first metacarpal and into the extensor mechanism of the thumb (Ch. VII).

Extensor Pollicis Brevis and Abductor Pollicis Longus

The tendons of the extensor pollicis brevis and of the abductor pollicis longus have an oblique course in the distal forearm which carries them across the tendons of the extensors carpi radialis (Fig. 71). They usually traverse the styloid process of the radius in a common sheath (Fig. 62), but occasionally the extensor pollicis brevis may have a separate sheath.

The extensor pollicis brevis tendon continues distally along the dorsal surface of the first metacarpal to join the extensor pollicis longus tendon, forming the extensor mechanism of the thumb. (See Ch. VII.)

The tendon of the abductor pollicis longus is much larger than the tendon of the extensor pollicis brevis. The abductor tendon lies palmar to the extensor tendon. The abductor pollicis longus is a powerful abductor of the first metacarpal and is supplied by the radial nerve.*

Radial Artery

In the distal volar forearm the radial artery courses along the lateral edge of the volar surface

---

* The abductor pollicis longus frequently has one or more aberrant tendons. The aberrant tendons may insert into the flexor retinaculum, the trapezoid, the abductor pollicis brevis (Fig. 18) and/or into the opponens pollicis.

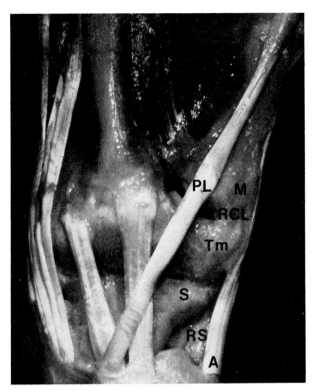

Figure 63. Floor of the snuffbox. A = abductor pollicis longus; M = first metacarpal; PL = extensor pollicis longus; RS = radial styloid process; S = scaphoid; Tm = trapezium.

of the radius. After reaching the tip of the radial styloid process, the artery angles dorsally and passes beneath the tendons of the abductor pollicis longus and of the extensor pollicis brevis. Traversing obliquely across the radial side of the hand, it continues beneath the tendon of the extensor pollicis longus and out over the first dorsal interosseous muscle.

During its course over the radial side of the hand, the radial artery gives off a dorsal carpal branch to the dorsal carpal network (Fig. 74), as well as giving off the first dorsal metacarpal artery (Fig. 75).

Thereafter, the radial artery passes between the two heads of the first dorsal interosseous muscle to enter the palm. Here it anastamoses with the deep branch of the ulnar artery to form the deep palmar arch (Ch. II). Occasionally the radial artery lies superficial to the tendons as it crosses the radial side of the hand (Fig. 66).

Extensors Carpi Radialis Longus and Brevis

Emerging from beneath the tendons of the abductor pollicis longus and of the extensor pollicis brevis, the tendon of the extensor carpi radialis

longus passes radial to Lister's tubercle. Continuing distally in a straight line, the extensor carpi radialis longus passes beneath the tendon of the extensor pollicis longus and inserts into the base of the second metacarpal (Fig. 63). The extensor carpi radialis brevis traverses a parallel course along the medial side of the extensor carpi radialis longus, inserting into the base of the third metacarpal.

Radial Collateral Ligament

The radial collateral ligament is the strongest and the most discrete of the carpal ligaments (Fig. 63). It connects the radial margins of the first metacarpal, the trapezium, and the scaphoid to the tip of the radial styloid process. These structures lie deep to the radial artery and to the tendons. The carpal joints are discussed in Chapters III and X.

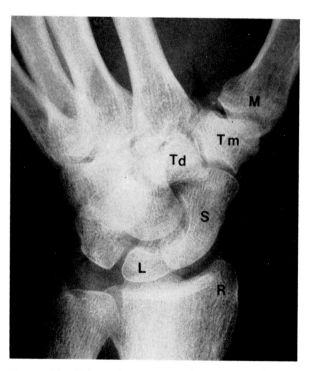

Figure 64. Oblique X-ray view of wrist. L = lunate; M = first metacarpal; R = radial styloid process; S = scaphoid; Td = trapezoid; Tm = trapezium.

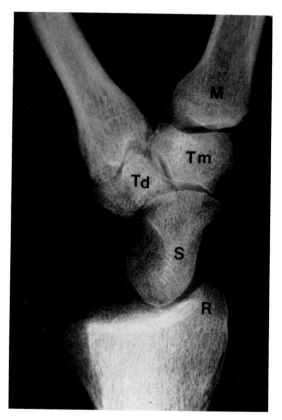

Figure 65. Oblique X-ray of radial side of hand, special preparation. M = first metacarpal; R = radial styloid process; Td = trapezoid; Tm = trapezium.

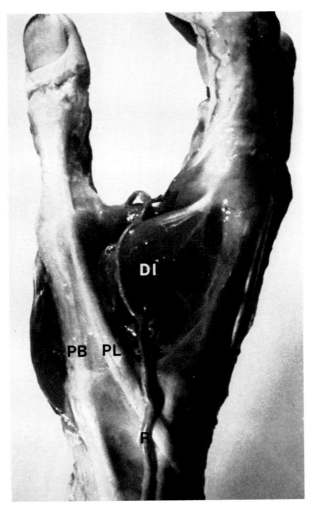

Figure 66. Aberrant radial artery. Note artery is superficial to tendons. DI = first dorsal interosseous muscle; PB = extensor pollicis brevis; PL = extensor pollicis longus; R = radial artery.

CHAPTER VI

DORSUM OF THE HAND

THE SKIN OF THE DORSUM OF THE HAND is thin and highly elastic. The layer of subcutaneous fat is thin. From the surgeon's pragmatic point of view, the skin and the subcutaneous fat form a single structure, which is only loosely tethered to the underlying structures. The thinness of the skin combined with this loose tethering increases the mobility of the dorsal skin and accommodates the simultaneous acute flexion of the wrist and of all the finger joints. The skin of most of the dorsum of the hand has an abundant blood supply. How-ever, the skin directly overlying the metacarpal heads of the four fingers has a sparse blood supply.

Dorsal Venous Network

The next layer of tissue, the filmy, transparent areolar tissue which covers the extensor retinaculum and the dorsal fascia of the hand, is easily overlooked. The dorsal venous network courses between the subcutaneous fat and the areolar layer (Fig. 67). The veins are loosely attached to both

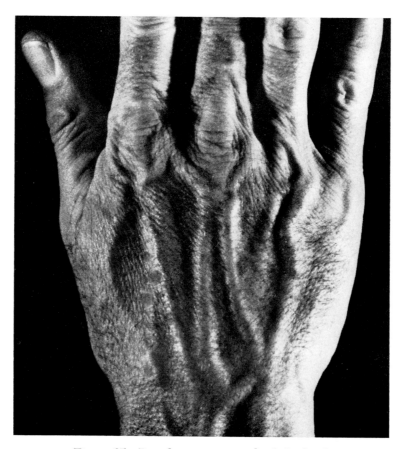

Figure 67. Dorsal venous network of the hand.

the overlying fat and to the underlying areolar layer and are easily separated from both. The primary venous drainage of the fingers is through these dorsal veins.*

The dorsal cutaneous sensory branches of the radial and ulnar nerves lie deep to the dorsal venous network and superficial to the filmy areolar layer. Although fine branches of these nerves supply the overlying skin, the nerves themselves are primarily attached to the underlying areolar layer. When the skin-fat layer is elevated, the nerves remain attached to the underlying areola, where they are readily visible.

Radial Nerve

The superficial branch of the radial nerve arborizes just distal to the radial styloid process (Fig.

---

* Although the surgeon should not allow the integrity of two or three dorsal veins to preclude an adequate exposure, the dorsal veins should not be sacrificed unnecessarily, as the resulting venostasis predisposes to postoperative edema and may result in secondary fibrosis.

---

68). One branch courses distally along the radial side of the dorsum of the thumb. Other branches pass distally between the metacarpal heads, bifurcate, and pass along the adjacent dorsal surfaces of the respective digits. Although the nerves may extend beyond the interphalangeal joint of the thumb, they rarely extend beyond the middle phalanx of the fingers. Branches of the proper palmar digital nerves supply the distal portion of the dorsum of the fingers (Figs. 38 and 43).

Ulnar Nerve

The dorsal branch of the ulnar nerve usually leaves the main trunk just proximal to the distal ulna, angling up onto the dorsum of the wrist (Fig. 69). One branch of the nerve courses distally along the ulnar side of the dorsum of the little finger. Another branch usually passes out between the fourth and fifth metacarpal heads, bifurcates, and extends onto the adjacent dorsal surfaces of the ring and little fingers.

Figure 68. Superficial branch of radial nerve. Note that in this specimen a branch (large arrow) extends to fourth interdigital fold to supply adjacent surfaces of ring and little fingers. Small arrows = terminal branches; R = superficial branch radial nerve.

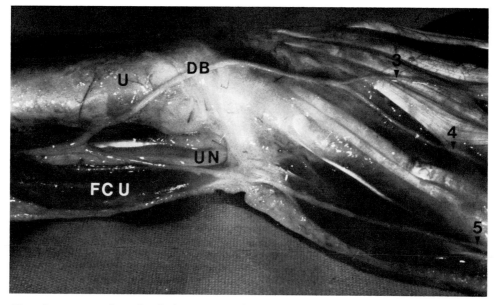

Figure 69. Dorsal cutaneous branch of ulnar nerve, ulnar aspect of hand. Note that in this specimen a branch extends to third interdigital fold to supply adjacent surfaces of long and ring fingers. DB = dorsal branch ulnar nerve; 3 = branch to third interdigital fold; 4 = branch to fourth fold; 5 = branch to ulnar side of little finger; FCU = flexor carpi ulnaris; U = distal end of ulna; UN = ulnar nerve.

The sensory distribution of the ulnar and radial nerves varies considerably. The ulnar nerve is classically described as innervating the skin of the little finger and the adjacent surface of the ring finger, along with a corresponding portion of the hand. The radial nerve is shown supplying sensation to the dorsum of the rest of the hand, and to the proximal two thirds of the radial half of the ring finger and of the long and index fingers and the thumb. Although this pattern is present in most instances, the radial nerve's territory may extend across the fourth metacarpal and onto the fourth interdigital fold (Fig. 68). Conversely, the ulnar nerve may send a branch across the fourth metacarpal and onto the third interdigital fold (Fig. 67).

Extensor Retinaculum

Deep to the nerves, and clearly visible through the filmy, transparent layer of areolar tissue, lies the dorsal retinacular system of the hand (Fig. 70). As the antebrachial fascia of the forearm extends distally toward the hand, it becomes thickened and opaque. This thickened fascial layer is the extensor retinaculum. It forms the roof of the six fibro-osseous tunnels through which pass the extensor tendons of the wrist and of the digits.

EXTENSOR TENDONS
Extensor Carpi Ulnaris

The tendon of the extensor carpi ulnaris is the most medial of the extensor tendons. It lies in the radial groove of the distal ulna and inserts into the proximal end of the fifth metacarpal at the ulnar tubercle (Fig. 71). As the tendon lies close to the axis of flexion of the wrist, the extensor carpi ulnaris is more effective as an ulnar deviator than as an extensor.

Extensor Digiti Minimi

Lying radial to the extensor carpi ulnaris, the twin tendons of the extensor digiti minimi course through a tunnel which is rigidly fixed to the radius. Occasionally this muscle contributes a separate tendon slip to the ring finger.

Extensor Digitorum

The extensor digitorum occupies the tunnel radially adjacent to that of the extensor digiti minimi. The extensor digitorum has a variable number of tendons. Although a single finger may receive multiple tendons from this muscle, they are rarely completely separated one from another, and they blend into a single tendon at the metacarpal head. Most often the extensor digitorum contributes a single tendon to the index, and two or more tendons each to the long and ring fingers, as well as a slip to the little finger. Obliquely placed cross-connections, called junctura, link together the tendons of the extensor digitorum over the metacarpal necks (Figs. 72 and 73).

Extensor Indicis

The tendon of the extensor indicis passes deep to those of the extensor digitorum in the same tunnel. Usually the extensor indicis has a single tendon which inserts into the extensor mechanism of the index finger on the ulnar side, whereas the extensor digitorum inserts on the radial side. Occasionally, the extensor indicis may have two tendons inserting into the index finger. In this instance, the tendons insert into the extensor mechanism of the index on either side of the tendon of the extensor digitorum (Fig. 72). Rarely, the extensor indicis may give off a separate tendon to the long finger.

Extensor Pollicis Longus

The tendon of the extensor pollicis longus parallels the extensor digitorum in the lower forearm. At Lister's tubercle, on the distal radius, the tendon courses radially toward the thumb (Fig. 71). Immediately distal to Lister's tubercle, the extensor pollicis passes over the tendons of the extensors carpi radialis longus and brevis.

Extensors Carpi Radialis

The tunnels for the radial wrist extensors lie to the radial side of Lister's tubercle (Fig. 71). The tendon of the extensor carpi radialis brevis passes over the distal radius close to the ulna and inserts into the base of the third metacarpal (Fig. 63). Because of its insertion into the third metacarpal, it is a pure extensor of the wrist. The extensor carpi radialis longus, having a more radial insertion into the base of the second metacarpal, is a powerful radial deviator of the wrist, as well as an effective wrist extensor.

Abductor Pollicis Longus and Extensor Pollicis Brevis

The abductor pollicis longus and the extensor pollicis brevis have origins which lie to the ulnar side of those of the extensors carpi radialis. The thumb abductor and extensor traverse the lower forearm in an obliquely radial direction, passing over the tendons of the extensors carpi radialis to gain positions along the lateral border of the radius. The two tendons extend distally over the radial styloid process through a fibro-osseous tunnel, the larger abductor tendon occupying the more lateral position. The smaller extensor pollicis brevis lies ulnar to the abductor. At the radial styloid these tendons pass through a rigid fibro-osseous tunnel.*

The extensor tendons have well-formed, true synovial sheaths beneath the extensor retinaculum (Fig. 93), and for varying distances distal thereto (Ch. X). All of the extensors of the wrist and of the digits, as well as the abductor pollicis longus, are supplied by the radial nerve.

Dorsal Carpal Arterial Network

Deep to the tendons, coursing over the ligamentous and capsular structures of the carpus, is the dorsal carpal network of arteries (Fig. 74). The network receives contributions from the dorsal carpal branches of the radial and ulnar arteries which traverse the wrist (Fig. 75). The dorsal branch of the anterior interosseous artery is the third component of the network. The vessels of the network are usually much smaller than those shown in traditional illustrations. Figure 75 illustrates one of the more prominent networks encountered by the authors. The dorsal carpal network supplies the carpal bones and gives rise to the dorsal metacarpal arteries which extend distally. The arteries pass between the metacarpal heads, bifurcating into the dorsal digital arteries (Ch. VIII).

Dorsal Ligaments

The ligamentous and capsular structures of the dorsum of the wrist form a relatively amorphous layer of tissue. The radial collateral ligament (Fig. 63), however, is a strong, discrete structure which connects the trapezium, the scaphoid, and the styloid process of the radius. Equally discrete is the ulnar collateral ligament (Fig. 110), which extends from the styloid process of the ulna to the pisiform and to the triangular. The re-

---

* Tenosynovitis (DeQuervain's disease) is common at this point. The extensor pollicis brevis will occasionally occupy a separate tunnel. This possibility should be borne in mind when operating for tenosynovitis in this area. Failure to decompress the extensor brevis, as well as the abductor pollicis longus, may result in incomplete relief of the patient's pain.

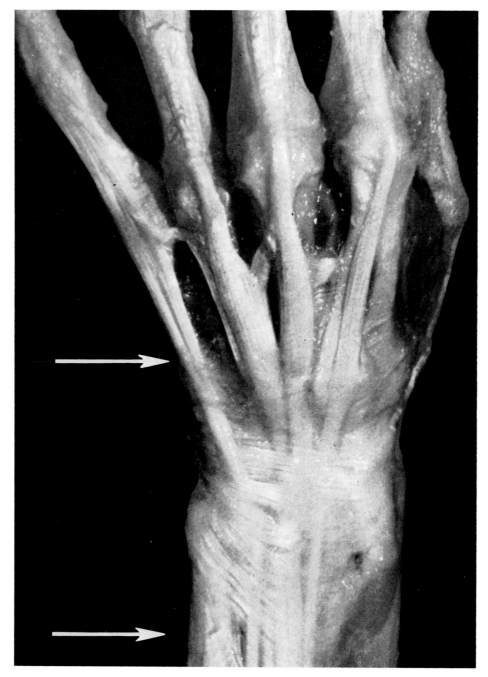

Figure 70. Extensor retinaculum. At proximal and distal margins (arrows) extensor retinaculum blends with translucent antebrachial fascia and transparent dorsal fascia of hand.

mainder of the numerous ligaments blend with the periosteum of the bones to form a single expanse of collagen (Fig. 76).*

* The surgeon will find little resemblance between what he sees in the operating room and the drawings of the anatomical treatises. So featureless are the dorsal capsular and ligamentous structures at surgery that it is difficult to identify the underlying joints.

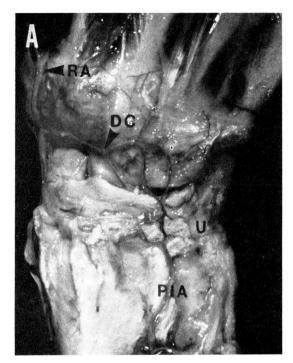

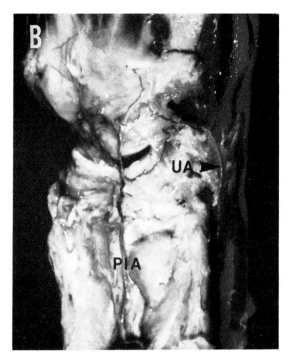

Figure 74. Dorsal carpal arterial network. A. dorsal view. B. dorso-ulnar view. DC = dorsal carpal branch of radial artery; PIA = posterior interosseous artery; R = radial styloid; RA = radial artery; U = ulna; UA = ulnar artery.

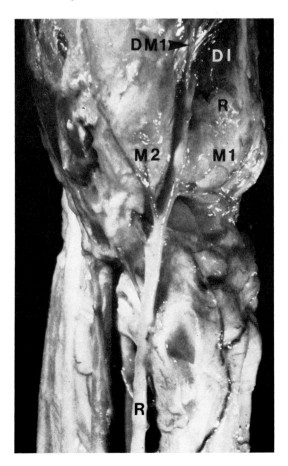

Figure 75. Radial artery, radial border of hand. Tendons removed. Note artery passing between heads of of first interosseous muscle to enter palm. DMI = first dorsal metacarpal artery; M1 = base of first metacarpal; M2 = base of second metacarpal; R = radial artery.

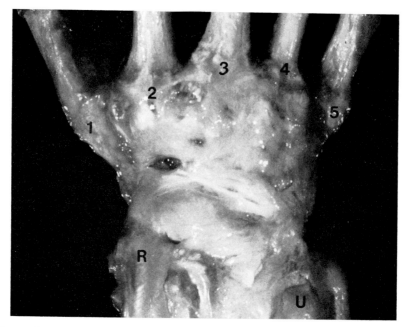

Figure 76. Dorsal carpal capsule and ligaments. An amorphus expanse of collagen. 1, 2, 3, 4, 5 = bases of metacarpals 1, 2, 3, 4, 5; R = radial styloid process; U = distal end ulna.

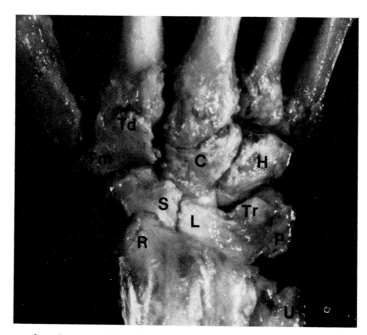

Figure 77. Carpal bones, dorsal aspect. Capsular and ligamentous tissues removed. Note that little articular cartilage is exposed when wrist in neutral extension. C = capitate; H = hamate; L = lunate; P = pisiform; R = radial styloid; S = scaphoid; Td = trapezoid; Tm = trapezium; Tr = triangular; U = ulnar styloid.

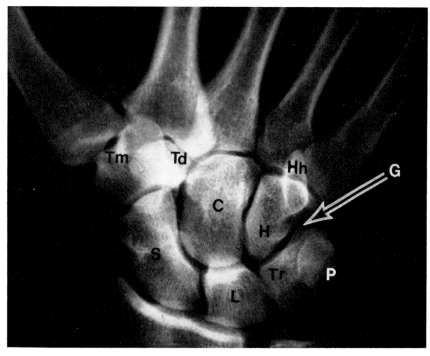

Figure 78. X-ray of carpal bones. Note space between hook of hamate and pisiform, called Guyon's canal (arrow). G = Guyon's canal; C = capitate; H = hamate; Hh = hook of hamate; L = lunate; P = pisiform; S = scaphoid; Td = trapezoid; Tm = trapezium; Tr = triangular.

CHAPTER VII

DORSUM OF THE THUMB

THE SKIN OF THE DORSUM OF THE THUMB is thin and moderately elastic. The subcutaneous fat is usually thin. The skin and the subcutaneous fat are firmly adherent to each other, but are tethered loosely to underlying structures. A network of veins beneath the skin and fat provide the principal venous drainage of the thumb.

Dorsal Digital Arteries

The dorsal digital arteries of the thumb lie on the extensor mechanism deep to the veins. Branches of the first dorsal metacarpal artery (Fig. 79), the dorsal arteries course along the dorsum of the thumb to the distal phalanx. These arteries supply the entire dorsal surface of the thumb. (In contrast, the dorsal digital arteries rarely extend much beyond the proximal phalanges.)*

Dorsal Digital Nerves

The dorsal digital nerves of the thumb (Fig. 80) are branches of the radial nerve. Usually they terminate before reaching the subungual area. The distalmost portion of the dorsum of the thumb is supplied by branches of the proper digital nerves. The proper palmar digital nerves of the thumb are terminal branches of the median nerve.

EXTENSOR MECHANISM

Extensor Pollicis Longus

The extensor mechanism of the thumb begins just proximal to the metacarpophalangeal joint, at the site of convergence of the tendons of the ex-

* This independent dorsal blood supply makes feasible the Moberg slide operation (M. A. Posner and R. J. Smith, "The Advancement Pedicle Flap for Thumb Injuries." *Journal of Bone and Joint Surgery*, Vol. 53A [1971], p. 1618.) In this procedure, the volar skin, along with the subcutaneous tissue including both neurovascular bundles, is advanced distally to cover a soft tissue defect on the volar surface of the distal phalanx of the thumb.

tensor pollicis longus and of the extensor pollicis brevis (Fig. 80). Although these tendons remain separate entities as they continue distally, they are intimately attached to one another.

The extensor pollicis longus inserts into the base of the distal phalanx and is therefore a constant extensor of the interphalangeal joint.

Extensor Pollicis Brevis

The tendon of the extensor pollicis brevis varies in length. The deep portion of the tendon inserts into the base of the proximal phalanx and extends the metacarpophalangeal joint. The superficial fibers of the tendon may continue distally for a variable distance as a component of the extensor mechanism. In some individuals fibers of the extensor pollicis brevis extend distally to insert into the distal phalanx (Fig. 79). In those individuals the extensor pollicis brevis extends the interphalangeal joint as well as the metacarpophalangeal joint.

Role of the Thenar Muscles

In addition to their primary functions, all of the intrinsic muscles of the thenar group except the opponens pollicis have secondary roles. The primary role of the abductor pollicis brevis is to oppose the thumb (Ch. II), and its secondary role is to extend the interphalangeal joint of the thumb. The primary function of the flexor pollicis brevis is flexion of the first metacarpal and of the metacarpophalangeal joint of the thumb. Its secondary role is extension of the terminal joint of the thumb. The adductor pollicis is primarily an adductor of the first metacarpal and a flexor of the metacarpophalangeal joint of the thumb. Secondarily, the adductor pollicis is an extensor of the interphalangeal joint of the thumb.

The abductor pollicis brevis, the flexor pollicis brevis, and the adductor pollicis have secondary roles as extensors of the distal joint of the thumb.

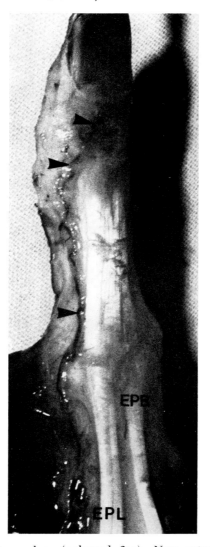

Figure 79. Thumb dorsal digital artery, ulnar (enlarged 2×). Note extensor pollicis brevis fibers extend to distal phalanx in this specimen. Arrows = ulnar side dorsal artery; EPB = extensor pollicis brevis; EPL = extensor pollicis longus.

because they have secondary insertions into the extensor mechanisms of the thumb. The opponens pollicis has no such secondary role because it inserts exclusively along the shaft of the first metacarpal.

The secondary tendons of insertion of these three muscles vary widely in size. The adductor pollicis, for example, may have a prominent tendon slip which extends from the muscle belly onto the dorsum of the thumb where it blends with the fibers of the extensor pollicis longus (Fig. 81). In other instances, the fibers which extend from the adductor pollicis to the extensor mechanism are thinner and their origins are more subtle.

The principal insertion of the abductor pollicis brevis is by a distinct tendon into the lateral capsule of the metacarpophalangeal joint of the thumb (Fig. 37). Fibers of this tendon continue distally and dorsally to contribute to the extensor mechanism. The flexor pollicis brevis inserts primarily into the radial sesmoid of the metacarpophalangeal joint. Secondary slips insert into the extensor mechanism as well as into the proximal phalanx.

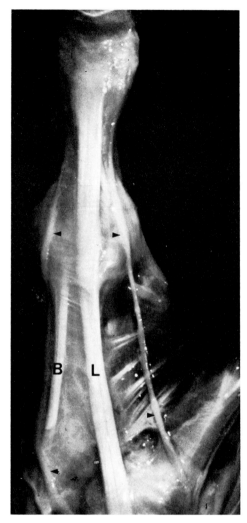

Figure 80. Dorsal nerves of the thumb (enlarged 2×). Arrows = dorsal nerves; B = extensor pollicis brevis; L = extensor pollicis longus. Note that the nerves extend well out onto the thumb.

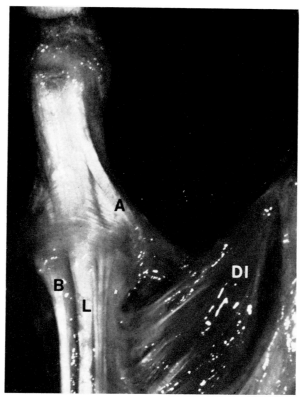

Figure 81. Extensor mechanism of the thumb (enlarged 2×). A = tendon slip from adductor pollicis into extensor mechanism; B = extensor pollicis brevis; DI = first dorsal interosseous; L = extensor pollicis longus.

CHAPTER VIII

DORSUM OF THE FINGERS

THE SKIN OF THE DORSUM of the fingers is thinner and more elastic than on the volar aspect. The subcutaneous fat is sparse and usually thin. The skin and subcutaneous fat are firmly bonded to each other but are only loosely bound to underlying structures. A network of veins which provides the principal venous drainage of the finger lies beneath the skin and fat.

The Nail

The fingernail covers the distal three quarters of the dorsum of the distal phalanx (Fig. 82). The proximal portion of the nail is covered by skin and eponychium. The proximal portion of the nail plate, covered by skin, overlies the germinal layer of the nail bed. Paronychium folds over and covers the lateral margins of the nail. The distal nail margin is not covered. The nail aids in stabilizing the skin and subcutaneous tissue of the fingertip. The nail plate is composed of keratin and is a derivative of the skin.

Dorsal Arterial Supply of the Finger

The dorsal digital arteries arise from the dorsal metacarpal arteries. The dorsal digital arteries of the fingers do not normally extend distally beyond the proximal interphalangeal joint. Many anatomical texts describe dorsal arteries extending out to the fingertips. The authors have been unable to demonstrate such arteries. Furthermore, other investigators have been unable to demonstrate dorsal arteries extending to the fingertips, regardless of whether they used radiographic techniques or employed more classical anatomical methods.[5, 6, 7]

The dorsum of the distal portion of the fingers is supplied by branches of the proper palmar digital arteries (Fig. 83). The entire dorsum of the thumb, however, has an independent blood supply derived from the first dorsal metacarpal artery.[8]

Dorsal Nerve Supply of the Finger

The dorsum of the proximal two thirds of each finger is supplied by a pair of dorsal digital nerves. These nerves arise from the radial and ulnar nerves on the dorsum of the hand (Ch. VI). The dorsal digital nerves course dorsolaterally on each side of each finger.

The distal third of the finger is supplied by dorsal branches of the proper digital nerves. These branches arise at the level of the proximal phalanx and run an oblique course onto the dorsum of the finger (Figs. 39, 43).

EXTENSOR MECHANISM

The extensor mechanism is a triangular structure covering most of the dorsum of the finger. Proximally the base of the triangle is draped over the dorsum and the sides of the metacarpal head. The apex of the triangle inserts into the distal phalanx. The extensor mechanism receives contributions from four muscles. In addition to the extrinsic extensor tendon, the mechanism receives one inter-

---

[5] Michael H. Flint, "Some Observations on the Vascular Supply of the Nail Bed and Terminal Segments of the Finger," *British Journal of Plastic Surgery*, Vol. 8 (1955), p. 186.

[6] Michael Flint and Stewart H. Harrison, "A Local Neurovascular Flap to Repair Loss of the Digital Pulp," *British Journal of Plastic Surgery*, Vol. 18 (1965), p. 156.

[7] J. H. Levame, C. Otero, and G. Berdugo, "Vascularization Arterielle des Teguments de la Face Dorsale de la Main et des Doits," *Annales de Chirurgie Plastique (Paris)*, Vol. 12 (1967), p. 316.

[8] Martin A. Posner and Richard J. Smith, "The Advancement of Pedicle Flap for Thumb Injuries," *Journal of Bone and Joint Surgery; American Volume (Boston)*, Vol. 53 (1971), p. 1618.

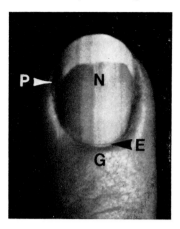

Figure 82. The nail. E = eponychium; G = region of skin-covered germinal layer of nail bed; N = nail plate; P = paronychium.

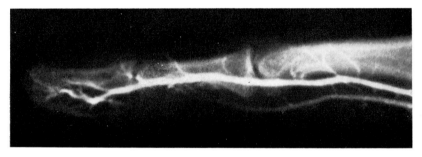

Figure 83. Arteriogram of the finger, lateral view. Note that although the dye was injected into the brachial artery of a living subject, no dorsal digital arteries are seen.

osseous tendon on each side, as well as a lumbrical tendon only on the radial side.*

The internal structure of the extensor mechanism is comprised of an array of diverging, converging, blending and decussating fibers (Figs. 84, 85, 86 and 87). At first glance, the pattern of fibers may appear haphazard, but in fact the arrangement is remarkably consistent. The sagittal band and an insertional slip into the proximal phalanx are consistent features of the extensor mechanism at the metacarpophalangeal joint. In the region of the proximal interphalangeal joint the mechanism is composed of a central slip which

---

* The tendinous contributions are not identical for each finger. The index finger has two extrinsic extensors (extensor communis and extensor indicis), a dorsal interosseous on the radial side, and a volar interosseous on the ulnar side. The long finger receives dorsal interosseous tendons on both the radial and the ulnar sides. The ring finger has a volar interosseous on the radial side and a dorsal interosseous on the ulnar side. The little finger has a volar interosseous on the radial side, and on the ulnar side the abductor digiti minimi acts in lieu of a fifth dorsal interosseous.

inserts into the middle phalanx and two lateral bands. The retinacular ligaments[9] extend palmarward from the lateral bands. Distal to the central slip lies the triangular ligament, and finally the decussation of the lateral bands just proximal to the insertion of the extensor mechanism into the distal phalanx.

Sagittal Band

The sagittal band covers the extensor tendon and the underlying capsule at the metacarpophalangeal joint. The sagittal band is composed of transversely oriented fibers which blend with the fibers of the extensor tendon. The band extends palmarward on each side of the metacarpophalangeal joint to insert into the deep transverse metacarpal ligament and into the fibrous tendon sheath. The proximal border of the sagittal band is distinct, but distally the fibers become indistinct.

As the extensor mechanism crosses the meta-

---

[9] Lee W. Milford, Jr., *Retaining Ligaments of the Digits of the Hand. Gross and Microscopic Anatomic Study* (Philadelphia, W. B. Saunders, 1968).

Figures 84, 85, 86, 87. Extensor mechanism of the finger, dorsal views. Figure 84. Extensor mechanism *in situ*. Figure 85. Intrinsic muscles detached proximally and brought up into plane of the dorsum. Figure 86. Extensor mechanism detached from finger and spread flat. Figure 87. Detached mechanism transilluminated. CS = central slip; D = decussation of lateral bands; DIP = distal interphalangeal joint; E = extrinsic extensor tendon; EF = extrinsic extensor fibers contributing to lateral band; IM = interosseous muscles; IND = insertion into distal phalanx; INM = insertion into middle phalanx; L = lumbrical; LB = lateral band; MP = metacarpophalangeal joint; PIP = proximal interphalangeal joint; R = ray fibers from intrinsic into central slip; S = sagittal band; TL = triangular ligament.

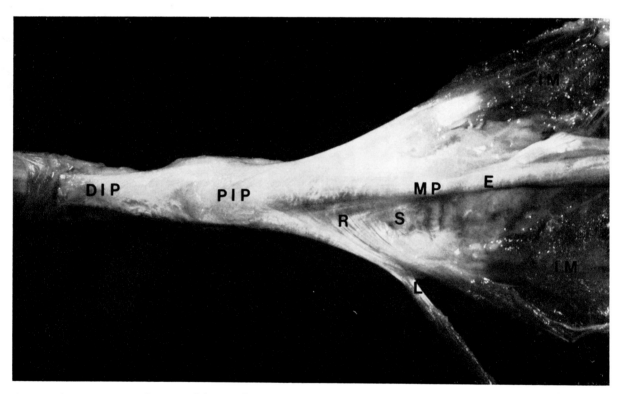

Figure 85. Extensor mechanism of finger, dorsal view. CS = central slip; E = extrinsic extensor tendon; IM = interosseous muscle; LB = lateral band.

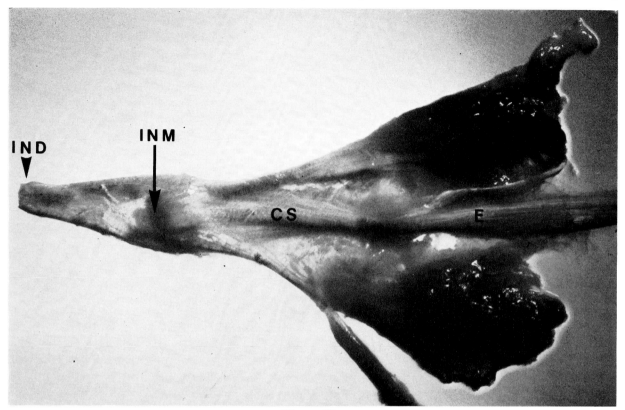

Figure 86.

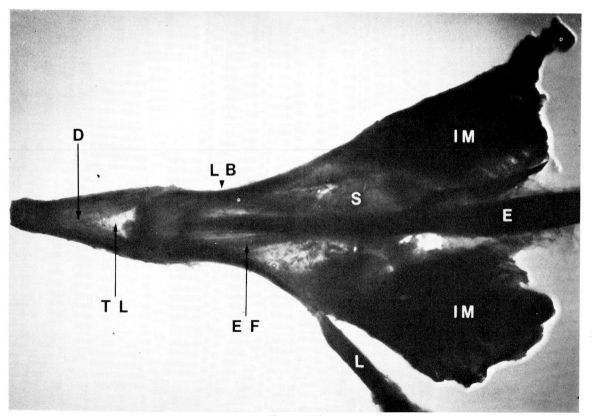

Figure 87.

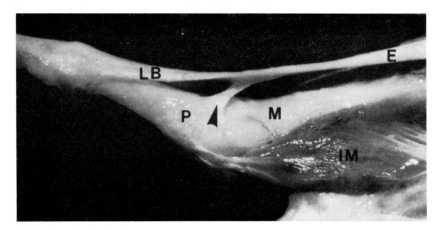

Figure 88. Extensor mechanism insertion into proximal phalanx. Lateral portion of extensor mechanism including sagittal band removed. Arrow = insertion of extensor mechanism into base proximal phalanx; E = extrinsic extensor tendon; IM = interosseous muscle; LB = lateral band; M = metacarpal; P= proximal phalanx.

carpophalangeal joint, the mechanism gives off a slip to the proximal phalanx. This slip inserts into the proximal end of the proximal phalanx (Fig. 88), along with the capsule of the metacarpophalangeal joint. At the base of the proximal phalanx, the fibers of the extrinsic extensor tendon split into three main groups. The three groups are the two lateral bands and the central slip.

Central Slip

The middle and largest group of fibers, called the central slip, passes directly distally across the proximal interphalangeal joint, blends with the dorsal capsule of the joint and inserts broadly into the proximal end of the middle phalanx (Fig. 90).*

Dorsal Interosseous Muscles

The tendons of the interosseous muscles flank the metacarpophalangeal joint. The dorsal interosseous muscles usually have two tendons of insertion, one long and one short (Fig. 90). The short tendon passes between the sagittal band and

* At the proximal and distal interphalangeal joints, the extensor mechanism and the dorsal capsule are bonded together. Anatomically they comprise two separate layers. Grossly and practically, however, they may be considered to be a single layer. In this book they are so considered.

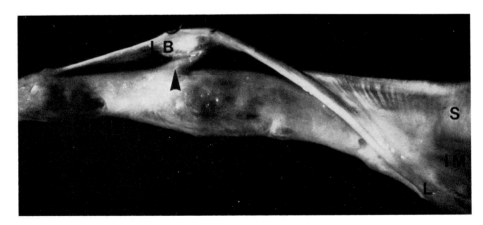

Figure 89. Insertion central slip into middle phalanx, lateral bands retracted dorsally (enlarged 1.5×). Note that lumbrical and interosseous are not separated by deep transverse metacarpal ligament because this illustration is of radial side of index finger. Arrow = insertion of central slip into base of middle phalanx; IM = interosseous muscle; L = lumbrical; LB = lateral band.

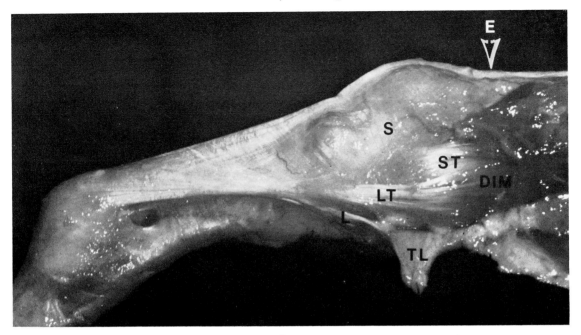

Figure 90. Extensor mechanism, radial side of long finger (enlarged 2×). DIM = dorsal interosseous muscle belly; E = extrinsic extensor tendon; L = lumbrical; LT = long tendon of interosseous inserting into extensor mechanism; S = sagittal band; ST = short tendon of interosseous inserting into proximal phalanx; TL = deep transverse metacarpal ligament (separating lumbrical from long tendon of interosseous).

the collateral ligament of the metacarpophalangeal joint to insert into the lateral aspect of the base of the proximal phalanx. The long tendon passes distally and dorsally superficial to the sagittal band, between the short tendon and the deep transverse metacarpal ligament. The fibers of the long tendon fan out over the extensor mechanism. Some of the fibers from the dorsal interosseous blend into the central slip, but the majority of the fibers merge into the lateral band.

Volar Interosseous Muscles

The volar interosseous muscles have constant insertions into the extensor mechanism but usually do not have a separate tendon inserting into the bone. Most of the fibers of the volar interosseous tendon merge into the lateral band, although a few fibers blend into the central slip.

Lumbrical Muscles

The lumbrical muscle enters each finger through the lumbrical canal, volar to the deep transverse metacarpal ligament (Fig. 90). The lumbrical tendon merges with the tendon of the interosseous on the radial side of the finger and thereby becomes a constituent of the extensor mechanism. From this point distally the fibers of the lumbrical

tendon cannot be distinguished from those of its companion interosseous tendon.

Lateral Bands

From their point of origin, the lateral bands diverge as they course distally. Skirting the proximal interphalangeal joint on either dorsolateral aspect, they then reverse course to converge over the dorsum of the middle phalanx. The converging lateral bands form the side of a triangle on the dorsum of the proximal portion of the middle phalanx. The insertion of the central slip occupies the base of the triangle. This triangular space is covered by the triangular ligament, a thin fibrous sheet which connects the two lateral bands.

The two converging lateral bands decussate and form a single tendon. This tendon inserts into the base of the distal phalanx.

Retinacular Ligaments

Each proximal interphalangeal joint has a pair of transverse retinacular ligaments, one on each side. The transverse retinacular ligament is a thin fibrous sheet covering the side of the joint. The ligament attaches the lateral margin of the lateral band of the extensor mechanism to the palmar ligament of the proximal interphalangeal joint

(Fig. 91). The transverse retinacular ligament is superficial to the oblique retinacular ligament and to the collateral ligament.

The oblique retinacular ligament is a delicate cord. Each proximal interphalangeal joint is spanned by a pair of oblique retinacular ligaments, one on each side (Fig. 92). Proximally, the ligament is attached to the distal border of the second pulley of the fibrous flexor tendon sheath. This attachment is in the midlateral line, at the junction of the pulley to the volar lateral ridge of the proximal phalanx. Distally the oblique retinacular ligament passes deep to the transverse retinacular ligament and superficial to the collateral ligament of the proximal interphalangeal joint. Distally the ligament blends into the margin of the lateral band.

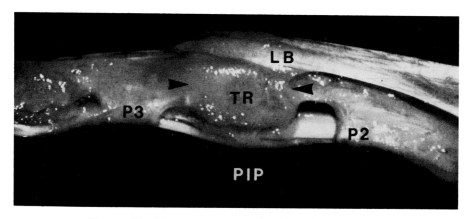

Figure 91. Transverse retinacular ligament (enlarged 2×). Note that the ligament extends from the lateral band to the flexor tendon sheath. Arrows denote ligament's proximal and distal margins. LB = lateral band; P2, P3 = second and third pulleys of the fibrous flexor tendon sheath; PIP = level of proximal interphalangeal joint; TR = transverse retinacular ligament.

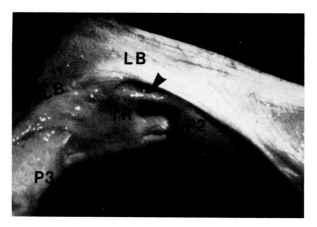

Figure 92. Oblique retinacular ligament (enlarged 2×). Note that ligament spans proximal interphalangeal joint deep to transverse retinacular ligament. Arrow = oblique retinacular ligament; LB = lateral band; P2, P3 = second and third pulleys of fibrous flexor tendon sheath; TR = transverse retinacular ligament.

CHAPTER IX

SPACES OF THE HAND

THE SOFT TISSUES OF THE HAND contain an extensive network of spaces. These spaces are of two types: true spaces and potential spaces. True spaces are found about the tendons, ie. the tendon sheaths and the subtendinous spaces. Potential spaces are present wherever two layers of tissue are only *loosely* connected to one another, e.g. the subcutaneous spaces of the fingers. When two such layers are split apart, as by the pressure created by the presence of pus or an injected foreign material, an actual space is created.

Some of the spaces connect constantly with adjacent spaces. In other instances, such communication does not exist under normal conditions but may easily be produced by the mechanical pressure of pus or of a foreign substance.

The network of true spaces and potential spaces extends from the digits to the forearm. There are communications between the dorsum of the hand and the palm, as well as between the radial and ulnar sides of the palm.*

Knavel's *Infections of the Hand*[10] is a definitive work on the role of anatomy in the propagation of infections of the hand. His beautiful studies of the spaces of the hand firmly established their role in the spread of infection. The authors have drawn freely from his work in the descriptions which follow. Knavel's terminology has been updated, but his anatomical concepts remain intact.

* Antibiotics have reduced the mortality rate from infections of the hand to the vanishing point, and they have simplified substantially their treatment. Nevertheless, an understanding of the role which the spaces of the hand play remains essential to the intelligent management of infections of the hand. Furthermore, recent industrial technologies which employ high pressure injection techniques have created a new category of injuries in which the spaces of the hand play an important part.

[10] Allen Knavel, *Infections of the Hand*, 6th ed. (Philadelphia, Lea & Febiger, 1953).

POTENTIAL SPACES

Subcutaneous Spaces of the Fingers

The subcutaneous spaces of the fingers begin distally at the proximal ends of the distal phalanges. These spaces extend circumferentially about the fingers except at the proximal radial side of the index finger and at the proximal ulnar side of the little finger. The volar portions of these potential spaces terminate at the level of the metacarpophalangeal joints. The dorsal portions of these spaces, however, continue proximally and merge over the metacarpophalangeal joints. The thumb has a similar potential subcutaneous space.

Dorsal Subcutaneous Space of the Hand

The merger of the dorsal subcutaneous spaces of the fingers at the level of the metacarpophalangeal joints forms the distal portion of the dorsal subcutaneous space of the hand. Knavel described the space as "an extensive area of loose connective tissue, without definite boundaries, which covers the entire dorsum of the hand." This potential space merges proximally into the subcutaneous tissues of the dorsum of the forearm.

TRUE SPACES

Dorsal Subtendinous Spaces of the Fingers

The dorsal subtendinous space of each finger begins at the level of the distal interphalangeal joint. The spaces lie between the extensor mechanisms of the fingers and the periosteum of the middle and proximal phalanges. The proximal boundary of each space is the fusion of the extensor mechanism into the proximal end of the proximal phalanx (Fig. 88). There is a similar space between the extensor mechanism of the thumb and the periosteum of the proximal phalanx.

Dorsal Subtendinous Space of the Hand

The dorsal subtendinous space of the hand (Fig. 73) lies between the extensor tendons of the fingers and the four underlying metacarpal bones. The dorsal surface of the space is formed by the sheaths, the junctura, and the paratenon of the extensor tendons. The palmar surface of the space is composed of the periosteum of the metacarpals and of the fascial covering of the interposed interosseous muscles. Usually the space ends in the hand and does not extend proximal to the wrist.

Extensor Tendon Sheaths

As the extensor tendons of the digits and the wrist pass through the six fibro-osseous tunnels beneath the extensor retinaculum (Ch. VI), they are enveloped by synovial sheaths (Fig. 93). Beginning at the lateral border of the radius, the tendons of the abductor pollicis longus and of the extensor pollicis brevis usually share a sheath. This single sheath extends distally to the metacarpophalangeal joint of the thumb.

The sheaths of the extensors carpi radialis longus and brevis are short. They terminate at the insertions of the tendons into the bases of the second and third metacarpals.

The sheath of the extensor pollicis longus is the longest of the extensor tendon sheaths. It begins several centimeters proximal to the wrist and extends to the metacarpophalangeal joint of the thumb.

The largest extensor tendon sheath accommodates both the tendons of the extensor digitorum and of the extensor indicis. This sheath terminates at the midmetacarpal level.

The sheath of the extensor digiti minimi is the sixth and most medially situated of the extensor tendon sheaths. This sheath begins proximal to the wrist and extends well out onto the fifth metacarpal.

Flexor Tendon Sheaths

The flexor tendons of each finger have an outer fibrous sheath and an inner synovial sheath. The volar and lateral walls of the outer sheath are formed by firm fibrous tissue. The dorsal wall is formed by the periosteum covering the phalanges and the palmar ligaments which bridge the joints. Along the shafts of the proximal and middle phalanges the volar and lateral walls consist of firm transverse bands of fibrous t'ssue (Fig. 43). At the interphalangeal joints, the volar and lateral walls are composed of annular and cruciate fibers. The fibrous sheaths hold the flexor tendons in close proximity to the underlying bones. The inner synovial sheaths form complete linings for the fibrous sheaths and provide a smooth gliding surface for the tendons. The synovial sheaths of the flexor tendons of the index, long and ring fingers terminate two or three centimeters proximal to the metacarpophalangeal joints (Fig. 94). The synovial sheath of the flexor tendon of the little finger usually continues proximally to become continuous with the ulnar bursa.

Ulnar Bursa

The ulnar bursa begins at the proximal termination of the tendon sheath of the little finger. From there, the bursa spreads out over the fourth metacarpal, as well as over the base of the third metacarpal. The ulnar bursa extends proximally beneath the flexor retinaculum and terminates three to five centimeters beyond it. Knavel found that "It does not surround the tendons completely but lies to the ulnar side of the superficial and deep tendons and envelops them as if they were invaginated into it from the radial side."

Radial Bursa

The radial bursa is the continuation of the digital portion of the tendon sheath of the flexor pollicis longus and of the continuous proximal expanded portion of the sheath (Fig. 94). This bursa begins distally at the insertion of the flexor pollicis longus. The bursa courses proximally over the volar surface of the proximal phalanx of the thumb, and then over the surfaces of the flexor pollicis brevis muscle and of the adductor pollicis muscle. Passing proximally through the carpal tunnel, the bursa ends three to five centimeters proximal to the flexor retinaculum. Communications are common between the radial bursa and the ulnar bursa at the wrist.

Mid-Palmar Space

The midpalmar space lies on the ulnar side of the palm (Fig. 95). The distal end of the space begins about two centimeters proximal to the

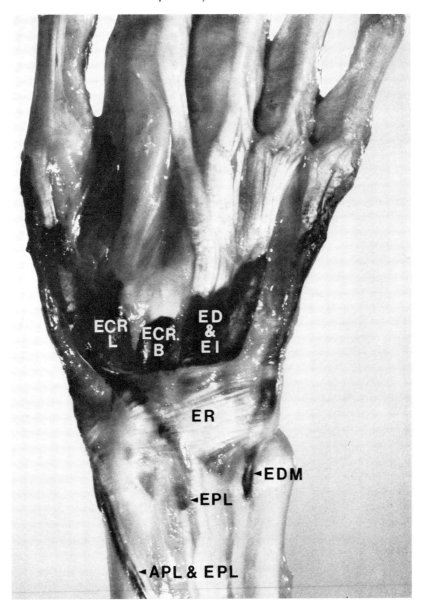

Figure 93. Extensor tendon sheaths, injected with black silastic. Note that the dense extensor retinaculum is opaque in some areas. APL and EPB = common sheath of abductor pollicis longus and extensor pollicis brevis; ECRB = extensor carpi radialis brevis; ECRL = extensor carpi radialis longus; ED and EI = common sheath of extensor digitorum and extensor indicis; EDM = extensor digiti minimi; EPL = extensor pollicis longus; ER = extensor retinaculum.

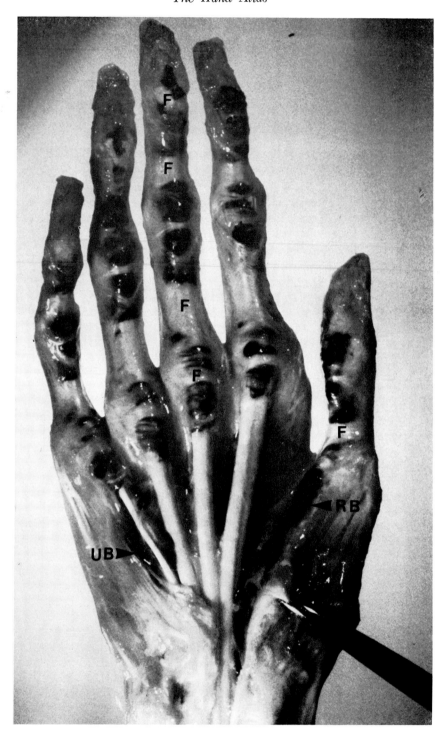

Figure 94. Flexor tendon sheaths, injected with black silastic. Note that flexor tendon sheath of little finger is continuous with ulnar bursa, sheath of flexor pollicis longus continuous with radial bursa. F = fibrous portions of flexor sheaths; RB = radial bursa; UB = ulnar bursa.

metacarpophalangeal joints. However, the space has diverticula which extend further distally along the lumbrical muscles of the ring and little fingers. The radial border of the space extends to the third metacarpal, where it is separated from the thenar space by a firm septum. Along the ulnar border, the space lies beneath the ulnar bursa from which it is separated by another fibrous septum. The midpalmar space is bounded dorsally by the palmar interosseous fascia and the ulnar border of the transverse head of the adductor pollicis muscle. The palmar boundary is a thin layer of fibrous tissue which separates the space from the flexor tendons and the lumbrical muscles of the long, ring and little fingers. The midpalmar space terminates proximally in a small isthmus which extends through the carpal tunnel into the forearm.

Thenar Space

The thenar space occupies the area between the thenar eminence and the third metacarpal (Fig. 96). It is bounded dorsally by the fascia covering the transverse head of the adductor pollicis muscle. Along its radial border, the thenar space is subcutaneous distally. Proximally the radial boundary is formed by the first metcarpal. The ulnar border of the space is a fibrous septum which ex-

tends the entire length of the third metacarpal shaft and separates the thenar space from the midpalmar space. Only a thin layer of fibrous tissue separates the palmar surface of the thenar space from the flexor tendons of the index finger. The distal extension of the spaces are diverticula which extend along the lumbrical muscles of the index and long fingers. Proximally, the thenar space ends at the proximal end of the first metacarpal.

Parona's Space

Parona's space extends from the carpal tunnel to the most distal point of attachment of the muscle bellies of the flexor digitorum profundus to the interosseous membrane (Figs. 97 and 98). The dorsal surface of the space is formed by the floor of the carpal tunnel and by the volar surface of the pronator quadratus muscle, as well as by the interosseous membrane. The radial and ulnar bursas and the flexors digitorum profundus cover the volar surface of Parona's space. The radial boundary of Parona's space in the forearm is formed by the deep fascia of the forearm which fuses with the periosteum of the radius. The ulnar boundary is the muscle and tendon of the flexor carpi ulnaris and the septum between that muscle and the flexor digitorum superficialis muscles.

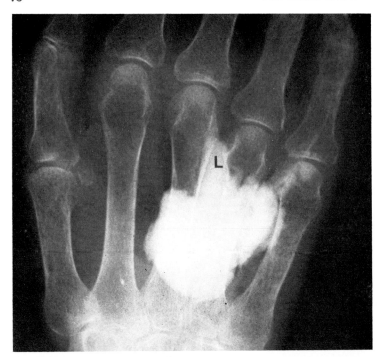

Figure 95. Mid-palmar space, injected with radiopaque material. L = diverticulum along third lumbrical.

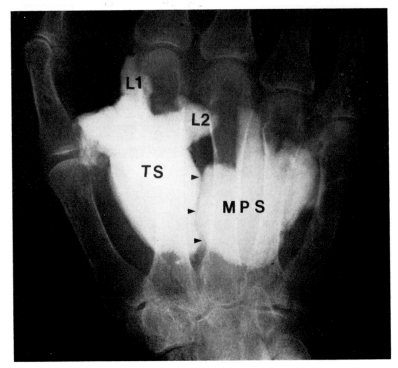

Figure 96. Thenar space, injected with radiopaque material. Arrows = septum separating thenar from mid-palmar space; L1, L2 = diverticula into first and second lumbrical canals; MPS = mid-palmar space; TS = thenar space.

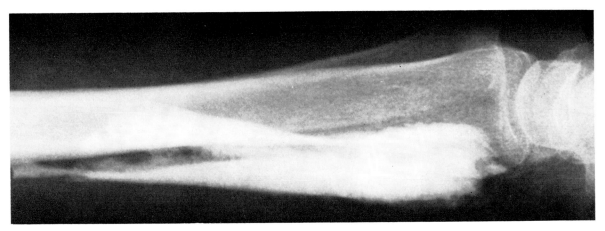

Figure 97. Parona's space, lateral view, injected with radiopaque media.

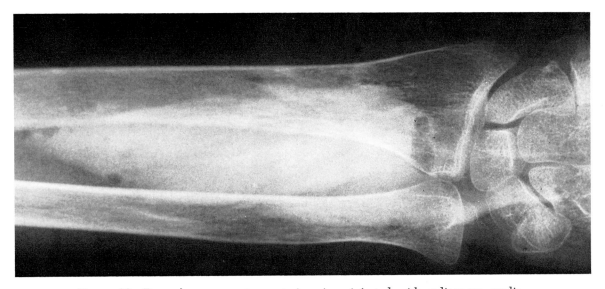

Figure 98. Parona's space, anteroposterior view, injected with radiopaque media.

CHAPTER X

WRIST MECHANICS

THREE TYPES OF MOTION are possible at the wrist. They are radial and ulnar deviation, flexion and extension, and pronation and supination. The joints involved are the radio-ulnar joint, the radiocarpal joints, the intercarpal joints, and the carpometacarpal joints. The contours of these articular surfaces are complex, reflecting the wide variety of combinations of gliding, angulatory and rotatory motions which they accommodate.

The amount of motion attributed to individual joints during various motions in the descriptions which follow are intended to be illustrative rather than definitive. These descriptions are based on radiographic observation made by the authors of the movements of major joints. This simplified analysis of wrist motion is at best a compromise. No attempt has been made at mathematical or engineering precision.

RADIAL AND ULNAR DEVIATION OF THE WRIST

Figure 99 is an anteroposterior X-ray view of a normal wrist in neutral position. A line drawn through the long axis of the radius coincides with a line drawn through the long axis of the third metacarpal. The long axis of the capitate is parallel to the long axes of the radius and of the third metacarpal, but lies slightly ulnarward. The plane of the ulnar border of the lunate in this specimen describes an angle of forty degrees with the axes of the radius and of the metacarpal.

As the wrist is deviated radially, the axis of the radius and that of the metacarpal intersect at an angle of fifteen degrees (Fig. 100). The angle between the plane of the ulnar border of the lunate and the axis of the radius has decreased five degrees. The capitate has rotated slightly on the lunate. The relationship between the third metacarpal and the capitate has not changed. The scaphoid has rotated in its long axis so that, viewed

in the anteroposterior plane of the wrist, it appears foreshortened some 12 per cent. This rotation permits the trapezium to impinge on the radial styloid and accommodates the radialward rotation of the capitate.

Figure 101 illustrates ulnar deviation of the wrist. The axis of the third metacarpal is ulnar deviated thirty-five degrees in relation to the longitudinal axis of the radius. Since it was radially deviated fifteen degrees in relationship to the same axis when the wrist was in radial deviation, the axis of the third metacarpal has shifted a total of fifty degrees in relationship to the longitudinal axis of the radius. The plane of the ulnar border of the lunate now forms an angle of twelve degrees with the long axis of the radius, twenty-eight degrees less than when the wrist was in neutral position.

The lunate and the scaphoid have shifted laterally on the radius. Simultaneously, the scaphoid has rotated in the opposite direction from that observed during radial deviation. Now, as viewed in the anteroposterior plane of the wrist, the scaphoid appears lengthened about 10 per cent. From ulnar deviation into radial deviation the scaphoid has appeared to increase its length 22 per cent. It would seem that if the scaphoid could neither move in relationship to the radius nor rotate in its long axis, little radial or ulnar deviation would be possible.

WRIST FLEXION AND EXTENSION

A specially prepared fresh specimen clearly illustrates by means of X-ray the basic motion which takes place during flexion and extension of the wrist. The skin, the subcutaneous tissue, and the extrinsic tendons were removed. For purposes of clarity, the ulna, the triangular, pisiform and hamate bones, as well as the fourth and fifth metacarpals, were removed (Fig. 103). The dissection

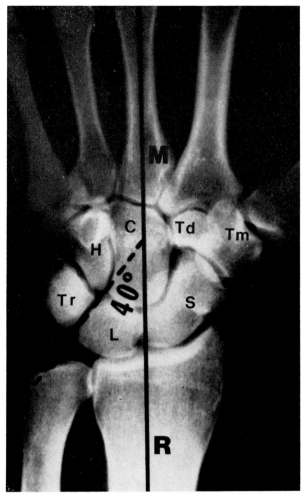

Figure 99. Anteroposterior X-ray of hand in neutral deviation. Axes of radius and third metacarpal coincide. Axis of ulnar border of capitate intersects radial axis at forty degree angle. C = capitate; H = hamate; L = lunate; P = pisiform; S = scaphoid; Td = trapezoid; Tm = trapezium; Tr = triangular; dashed line = axis of ulnar border of lunate.

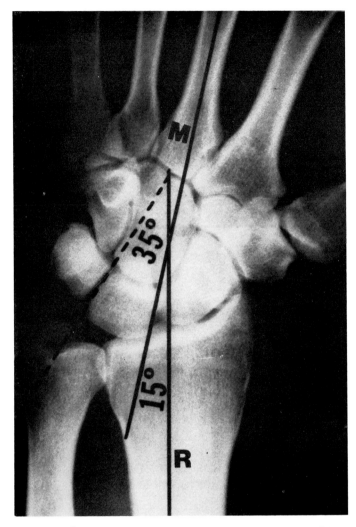

Figure 100. Anteroposterior view of wrist in maximum radial deviation. Scaphoid has rotated in its long axis, appearing foreshortened. Long axes of third metacarpal and radius form fifteen degree angle. Angle between long axis of radius and ulnar border of lunate has decreased five degrees. M = long axis of third metacarpal; R = long axis of radius; dashed line = axis of ulnar border of lunate.

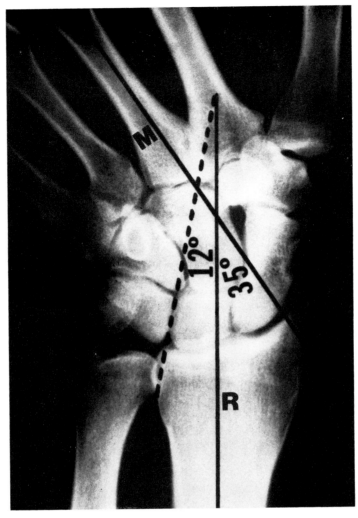

Figure 101. Anteroposterior view of wrist in maximum ulnar deviation. Scaphoid has rotated in its long axis, appearing lengthened. Axis of third metacarpal now ulnar deviated thirty-five degrees in relation to radial axis. Axis of ulnar border of lunate now forms twelve degree angle with radial axis. M = long axis of third metacarpal; R = long axis of radius; dashed line = axis of radius; dashed line = axis of ulnar border of lunate.

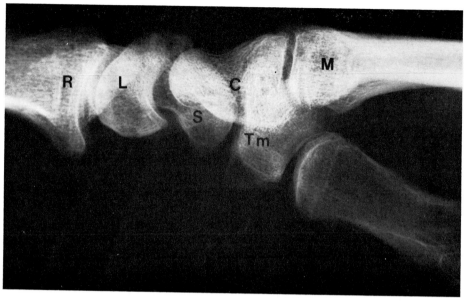

Figure 102. Lateral view of special preparation, neutral extension. Ulna, triangular, pisiform, hamate, fourth and fifth metacarpals removed. C = capitate; L = lunate; M = third metacarpal; R = radius; S = scaphoid; Tm = trapezium.

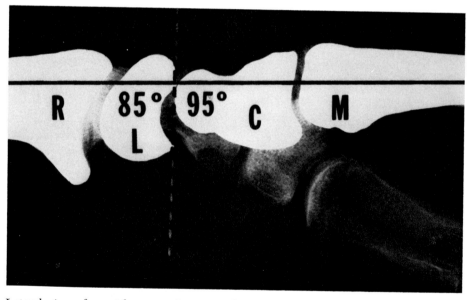

Figure 103. Lateral view of special preparation, neutral extension. Radius, lunate, capitate and third metacarpal have been silhouetted. Vertical axis of distal articular surface of lunate is almost perpendicular to coinciding axes of radius and capitate.

of the specimen may well have resulted in exaggerated mobility, but it cannot have decreased mobility.

The specimen is shown in the lateral plane in neutral extension in Figure 103. The radius, lunate, capitate and the third metacarpal have been silhouetted. The vertical axis of the distal articular surface of the lunate intersects the axis of the radius at an angle of eighty-five degrees. The axis of the lunate intersects the longitudinal axis of the capitate at an angle of ninety-five degrees.

When the wrist is flexed maximally (Fig. 104), the angle between the axis of the lunate and that of the radius is reduced to forty degrees (Fig.

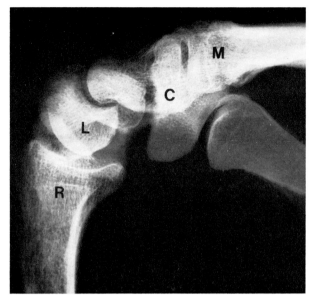

Figure 104. Lateral view of special preparation, maximum flexion.

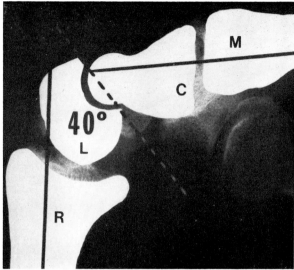

Figure 105. Wrist flexion. Angle between lunate axis and radial axis now forty degrees.

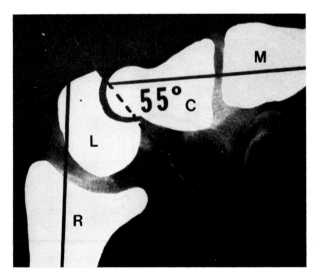

Figure 106. Wrist flexion. Angle between lunate axis and metacarpal axis now fifty-five degrees.

105). Simultaneously, the axis of the lunate now intersects that of the capitate at a fifty-five-degree angle (Fig. 106). Forty-five degrees of flexion have taken place at the radiocapitate joint and forty degrees at the lunocapitate joint, for a total of eighty-five degrees of wrist flexion.

As the wrist is maximally extended (Fig. 107), the radiolunate angle is increased to 115 degrees (Fig. 108). Simultaneously, the lunocapitate angle is increased to 130 degrees (Fig. 109). The radiolunate joint and the lunocapitate joint contributed,

respectively, thirty and thirty-five degrees of wrist extension. Total wrist extension is sixty-five degrees.

PRONATION AND SUPINATION OF THE WRIST

The distal articular surface of the radius is triangular in shape. The apex of the triangle is formed by the styloid process of the radius, and the base is formed by the concave ulnar notch.

The ulnar notch is covered by hyaline cartilage and receives the radial articulation of the ulna. The concave distal surface of the radius (Fig. 101) articulates with the scaphoid and the lunate. The ulna, however, does not articulate with the carpal bones. It is separated from them by the triangular fibrocartilage. This structure, two millimeters thick at its radial attachment, becomes four or five millimeters thick at its attachments to the base of the ulnar styloid process and the ulnar collateral ligament (Figs. 110 and 111).

To illustrate pronation and supination, X-rays were made of a normal wrist immobilized in a plaster thumb spica cast. The cast permitted pronation and supination, but eliminated both flexion and extension and radial and ulnar deviation. The essential motion at the wrist in pronation and supination is rotation of the radius about the ulna. The ulna lies more lateral to the radius in pronation than it does in supination. This minor shift is

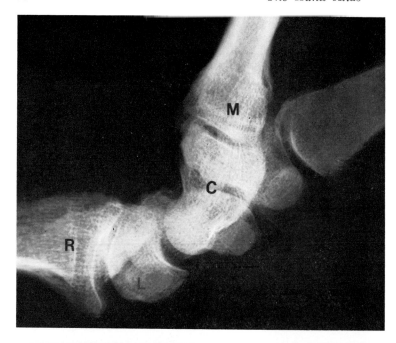

Figure 107. Lateral view of special preparation, maximum extension.

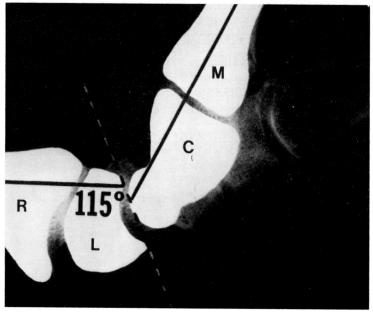

Figure 108. Wrist extension. Angle between lunate axis and radial axis now 115 degrees.

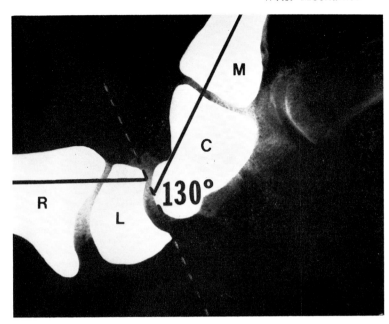

Figure 109. Wrist extension. Angle between lunate axis and metacarpal axis now 130 degrees.

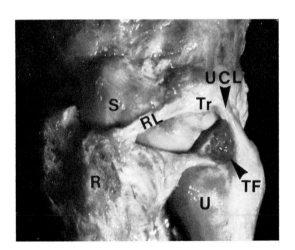

Figure 110. Radio-ulnar joint, dorsal aspect. Joint capsule and synovium removed. L = lunate; R = radius; RL = radiocarpal ligament; S = scaphoid; TF = triangular fibrocartilage; Tr = triangular bone; U = ulna; UCL = collateral ligament.

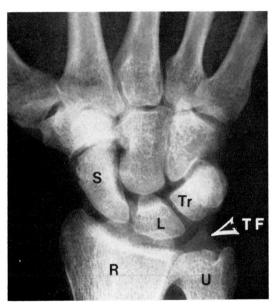

Figure 111. X-ray of specimen shown in Figure 110. L = lunate; R = radius; S = scaphoid; TF = triangular fibrocartilage; Tr = triangular bone; U = ulna.

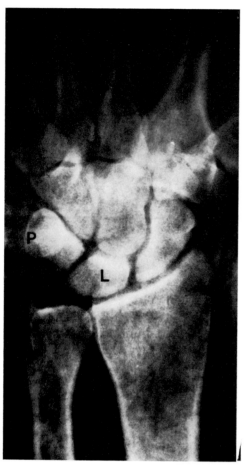

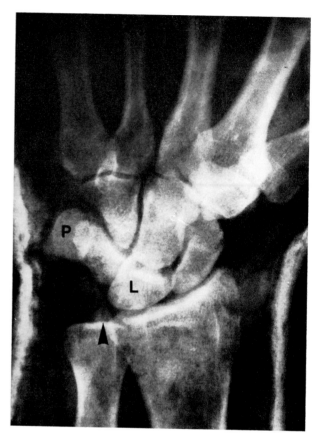

Figure 112. Wrist pronated. Ulnar styloid (arrow) is in line with pisiform, and only about 60 per cent of lunate articulates with radius. L = lunate; P = pisiform.

Figure 113. Wrist supinated. Lunate articulates almost exclusively with radius, ulnar styloid (arrow) is aligned with medial border of lunate. L = lunate; P = pisiform.

reflected in the degree of overlap of the distal ends of the radius and ulna, as seen in the antero-posterior plane. Additional changes occur at the radiocarpal joints. When the wrist is fully pronated, about 60 per cent of the proximal surface of the lunate articulates with the radius and 40 per cent with the triangular cartilage (Fig. 112). During supination, however, the lunate articulates completely with the radius (Fig. 113).

CARPOMETACARPAL JOINTS

Virtually no motion takes place between the capitate and the third metacarpal. There is slight motion between the trapezoid and the second metacarpal, and there is moderate mobility at the fourth carpometacarpal joint. The fifth carpometacarpal joint is the most mobile of the four, but even this joint permits only a few degrees of motion in each plane. In contrast, the trapezio-metacarpal joint of the thumb provides for a wide range of motion in all planes (Ch. III).

BIBLIOGRAPHY

Coleman, S. S., and Anson, B. J.: Arterial patterns in the hand based upon a study of 650 specimens. *Surg Gynecol Obstet, 113*:409, 1961.

Flint, M. H.: Some observations on the vascular supply of the nail bed and terminal segments of the finger. *Br J Plast Surg, 8*:186, 1955.

Flint, M. H., and Harrison, S. H.: A local neurovascular flap to repair loss of the digital pulp. *Br J Plast Surg, 18*:156, 1965.

Hartz, M.: The dermal papillae in the fingertip. *Plast Reconstr Surg, 45*:141, 1970.

Knavel, A.: *Infections of the Hand,* 6th ed. Philadelphia, Lea & Febiger, 1933.

Levame, J. H., Otero, C., and Berfugo, G.: Vascularization arterielle des téguments de la face dorsale de la main et des doits. *Ann Chir Plast, 12*:316, 1967.

Milford, L. W., Jr.: *Retaining Ligaments of the Digits of the Hand. Gross and Microscopic Anatomic Study.* Philadelphia, Saunders, 1968.

Posner, M. A., and Smith, R. J.: The advancement of pedicle flap for thumb injuries. *J. Bone Joint Surg [Am], 53*:1618, 1971.

SUGGESTED READING

Harris, C., Jr., and Rutledge, G. L., Jr.: The functional anatomy of the extensor mechanisms of the finger. *J Bone Joint Surg [Am], 54*:713, 1972.

Kaplan, E. B.: *Functional and Surgical Anatomy of the Hand,* 2nd ed. Philadelphia, Lippincott, 1965.

Moberg, E.: Aspects of sensation in reconstructive surgery of the upper limb. *J Bone Joint Surg [Am], 46*:817, 1964.

INDEX

91